# Interdisciplinarity in the Scholarly Life Cycle

Karin Bijsterveld • Aagje Swinnen
Editors

# Interdisciplinarity in the Scholarly Life Cycle

Learning by Example in Humanities and Social Science Research

palgrave
macmillan

*Editors*

Karin Bijsterveld
Department of Society Studies
Faculty of Arts & Social Sciences
Maastricht University
Maastricht, The Netherlands

Aagje Swinnen
Department of Literature and Art
Faculty of Arts & Social Sciences
Maastricht University
Maastricht, The Netherlands

ISBN 978-3-031-11107-5        ISBN 978-3-031-11108-2    (eBook)
https://doi.org/10.1007/978-3-031-11108-2

Cover illustration: © Eric Bleize

This Palgrave Macmillan imprint is published by the registered company Springer Nature Switzerland AG.
The registered company address is: Gewerbestrasse 11, 6330 Cham, Switzerland

# Acknowledgments

Academic books have their own scholarly life cycle. This volume acquired early intellectual and financial support from Thomas Conzelmann, vice-dean of research at the Faculty of Arts and Social Sciences (FASoS) of Maastricht University, our academic home. His successor Sally Wyatt shared his enthusiasm for the project and contributed a chapter to the book herself. The other members of the FASoS Board, Christine Neuhold and Giselle Bosse, underscored the significance of showing what interdisciplinary research means in practice, to students and scholars alike. We are very grateful for the funding they released for this book project.

One highlight in preparing the volume was a workshop in which artist Marte Hameleers and photographer Maaike Faas-Schauer invited us to stop writing for a moment and guided us in capturing our interdisciplinary work in artworks for this book. The workshop did not only make us see our scholarly work in a different light, but, at times, it also pushed that work in a new direction, as making brings understanding. For both Hameleers and Faas-Schauer, the workshop was their first experience with online teaching; it took place in the midst of the Covid-19 pandemic. In fact, the entire project ran during the pandemic years of 2020 and 2021. Despite the physical distance, Hameleers and Faas-Schauer were able to reach and touch us in the good sense of the word—many thanks! The contribution by Eric Bleize, photographer and FASoS scheduler, was touching in another way. While he considered it an honor to have one of his photos on the cover of this book, we feel honored that he gave permission to use his wonderful work of art, as another token of collaboration across disciplinary boundaries.

We would also like to thank Michele Faguet and Ton Brouwers for the swift and precise ways in which they corrected our English and their understanding for the differences in style that interdisciplinary work brings. Not only did we have their eyes to co-read our work but also those of the colleagues who spend precious time—even more precious in pandemic times—to blindly review our work. We appreciate their suggestions, as well as those by the contributors who commented on each other's and our own work.

Our first editor at Palgrave was Rebecca Wyde. Her thinking along with us on the title of the book was indispensable, as was her overall support for the book. It was a pleasure to work with her and with Sarah Hills, her successor at Palgrave. Hills knew solutions for every potential problem and took over some of the administrative burdens that come with publishing books. We felt surrounded by genuine professionals even though everything was "remote."

During the pandemic, "close" and "remote" acquired different meanings. To one of us—Karin—Rein de Wilde remained close as ever, and it was great to have such an attentive co-reader at home. In fact, the project helped to keep our academic life alive and, to use a Dutch word, *gezellig*. We, the editors, were both working in the nearly sealed-off rooms of our private offices—pale faces, a white glow of light against a dark background. Yet, talking about the book, and chatting about much more, brightened up our days. Academic life is now starting up again, and the book is out. On the desks and screens of others, we hope, to continue the life cycle of its own.

# Contents

# Notes on Contributors

**Karin Bijsterveld** is a historian and Professor of Science, Technology, and Modern Culture at Maastricht University. She has worked on the history of aging and on themes at the intersection of STS and sound studies. See, for instance, *Sonic Skills: Listening for Knowledge in Science, Medicine and Engineering* (Palgrave, 2019).

**Imogen Eve**, trained in the performing arts, worked as an artistic researcher at the Maastricht Centre for the Innovation of Classical Music (MCICM). From this collaborative research—which incorporated an ethnographic approach—Imogen developed and published a multi-genre and education-oriented text, *The Same but Differently* (2021). Her research interests include arts labor, education, storytelling, and literary ethnography.

**Elsje Fourie** is a member of the Globalization, Transnationalism, and Development (GTD) research group at Maastricht University. Her research focuses on the influence of East Asian approaches to national development on African modernist imaginaries. It has been published widely in academic journals and prominent news outlets.

**Kathleen Gregory, Paul Groth, Andrea Scharnhorst, and Sally Wyatt** The authors of this chapter worked together between 2017 and 2020 on the ReSEARCH project, funded by the Dutch Research Council. They have degrees in computer science, economics, education, library and information science, philosophy, physics, and STS from universities in Canada, England, Germany, the Netherlands, and the United States. They now work in the Netherlands and Austria but have prior experience in England, Germany, and the United States and have held visiting positions in many other countries. The study *Findable and Reusable? Data Discovery Practices in Research* forms the basis of this chapter. It draws on insights from information science and STS and uses multiple methods to understand how researchers discover data they might subsequently re-use in their own work.

**Ferenc Laczó** is Assistant Professor with tenure in History at Maastricht University. Laczó is the author, editor, or co-editor of eleven books. He studied, taught, or held fellowships in Basel, Berlin, Bielefeld, Budapest, Jena, Los Angeles, Utrecht, Vienna, and Washington, DC.

**Karin van Leeuwen** is Assistant Professor in History at Maastricht University. Her research focuses on the intersections of political and legal history. After a PhD on Dutch constitutional history, van Leeuwen published various articles on the history of EU law. Currently, she works on international law and the League of Nations.

**Harro van Lente** graduated in physics and philosophy in 1988. Since 2014, he is Professor of Science and Technology Studies at Maastricht University. He published extensively on the dynamics of emerging technologies, technology-society interaction, and the politics of knowledge production.

**Flora Lysen** is a post-doc researcher at the Society Studies Department of Maastricht University, investigating artificial intelligence in clinical decision-making. She was the first program coordinator of the Amsterdam Research Institute of the Arts and Sciences (ARIAS) and is currently a member of MERIAN (Maastricht Experimental Research In and Through the Arts Network).

**Valentina Mazzucato, Bilisuma Dito, and Karlijn Haagsman** Mazzucato, Dito, and Haagsman started working together in 2010. Mazzucato, Professor of Globalization and Development, worked in international organizations before getting a PhD from Wageningen University. After working at various universities, she joined Maastricht University where she headed the TCRAf-Eu project on transnational migrant families between Europe and Africa. Karlijn Haagsman was the first PhD student on the project. Haagsman was trained in Anthropology and Migration Sociology, respectively. Her mixed background fits the mixed-method interdisciplinary TCRAf-Eu project. She co-conducted one of the first surveys on transnational families and currently works on the mobility of migrant youth on the MO-TRAYL project. Bilisuma Dito joined the team one year later. She has a PhD in Development Economics (Erasumus University) and worked at Addis Ababa University. In the TCRAf-Eu project, Dito's work investigated well-being in households. She now works in the research project 'Wellbeing, Women, and Work in Ethiopia (3WE).'

**Jessica Mesman** is Associate Professor of Science and Technology Studies at Maastricht University. Her research interest centers on the role of implicit and informal knowledge in stabilizing complex medical practices. In order to do so, she has developed an 'exnovative' approach and frequently uses the method of video-reflexive ethnography.

**Peter Peters** is Associate Professor and Endowed Professor of Innovation of Classical Music at the Faculty of Arts and Social Sciences, Maastricht University. He is also director of the Maastricht Centre for the Innovation of Classical Music (MCICM), a collaboration between the South Netherlands Philharmonic, Zuyd University for Applied Sciences, and Maastricht University.

**Jos Roeden** studied clarinet at the Conservatories of Maastricht and The Hague, and Art and Business at Leiden University. Since 2014, he has been the artistic programmer of the South Netherlands Philharmonic. On behalf of the orchestra, he is also opera and classical music coordinator at Parkstad Limburg Theaters, Heerlen.

**Emilie Sitzia** holds a special chair as Professor of Word/Image at the University of Amsterdam (Fiep Westendorp Foundation) and is an associate professor at Maastricht University. She teaches interdisciplinary research methods, illustration history, cultural education, curatorship, and museology. Sitzia publishes regularly on art, literature, and museum studies topics.

**Paul Stephenson**  is an associate professor at the Department of Political Science, Maastricht University. He has published in academic journals including *Public Management Review* and *Modern and Contemporary France*. His research interests include EU institutions, transport and cohesion policy, and contemporary French politics and society.

**Aagje Swinnen**  is Professor of Aging Studies at Maastricht University. She has published on representations of aging in literature, photography, and film; art interventions in dementia care; and professional artists' understandings of late-life-creativity. Swinnen is co-founder of the European Network in Aging Studies and the journal *Age, Culture, Humanities*.

**Patricia de Vries**  is research professor at the Gerrit Rietveld Academy. Her work resides at the intersection of philosophy, art, and technology. She has published in *Big Data & Society*, *Rhizomes*, *Krisis*, and *Amsterdam Book Review* and has written essays for arts institutions in Eindhoven, Utrecht, Rome, and Shanghai.

**Joseph Wachelder** graduated as a theoretical physicist. He is now Associate Professor of Cultural History of Science and Technology in the Department of History, Maastricht University. Wachelder has published on the history of higher education, science shops, and toys. He co-edited *Materializing Memories: Dispositifs, Generations, Amateurs* (2018).

**Ties van de Werff** is a researcher at the Research Centre for Arts, Autonomy and the Public Sphere, and a teacher in the interdisciplinary bachelor iArts at Zuyd University of Applied Sciences, Maastricht. He was a postdoctoral researcher at the MCICM. He currently teaches and studies ethics of artistic research.

# List of Figures

## Spatial Rituals and Ritualized Space in Dutch Postwar Homes for the Elderly: Anthropology in History

## The Mysterious User of Research Data: Knitting Together Science and Technology Studies with Information and Computer Science

## Interdisciplinary Anticipations: Art-Science Collaboration at the Maastricht Brain Stimulation and Cognition Laboratory

## Doing Collaborative Research on Symphonic Orchestra Audiences: Interventionist Ethnography of Music Practices

## 'Doing' Teamwork as 'Doing' Family: Researching Transnational Migrant Families Through Interdisciplinary Collaboration

# Introduction

Karin Bijsterveld and Aagje Swinnen

## Spring Reading

In a 2021 essay about her personal intellectual development, Lorraine Daston recalls how one day in spring 1975, she chose a thin book from a stack of new publications in Harvard University's Widener Library. She was a first-year graduate student in the history of science program at the time and opted for a short book because she was babysitting that evening and did not expect to have much time to read. However, the boy she took care of went to sleep early, so she was able to read the book—Ian Hacking's *The Emergence of Probability: A Philosophical Study of Early Ideas about*

K. Bijsterveld (✉)
Department of Society Studies, Faculty of Arts & Social Sciences, Maastricht University, Maastricht, The Netherlands
e-mail: k.bijsterveld@maastrichtuniversity.nl

A. Swinnen
Department of Literature and Art, Faculty of Arts & Social Sciences, Maastricht University, Maastricht, The Netherlands
e-mail: a.swinnen@maastrichtuniversity.nl

K. Bijsterveld, A. Swinnen (eds.), *Interdisciplinarity in the Scholarly Life Cycle*,
https://doi.org/10.1007/978-3-031-11108-2_1

*Probability, Induction and Statistical Inference* (1975)—from cover to cover. Daston found it fascinating that Hacking used history to ask a genuinely new philosophical question: why had it taken so long to develop probability theories even though games of chance had been around for ages? Rather than attempt to unravel the conceptual chaos of probability theories, as earlier philosophers had done, Hacking's starting point was a research puzzle.

Reading Hacking's book showed Daston how one could pursue philosophy in a historical manner and history with a philosophical twist. Hacking's example of conceptually informed, yet radically empirical research also influenced Daston's own work in the history of science. It granted her the intellectual freedom to leave the trodden path of known methods for less clearly delineated ones and helped her to understand that the definition of the historical phenomena under study also requires explanation. There was an additional event, however, that was just as formative. Not long after finishing her PhD, she joined a research group cofounded by Hacking at the Centre for Interdisciplinary Research in Bielefeld, Germany. As the introductory meeting concluded, it was suggested that someone take notes. Rather than ask a junior scholar, Hacking offered to do it himself. This set an important precedent: his act embodied the "egalitarian tone" vital to "the coherence of the group" (Daston, 2021, p. 73), showing that it was possible for researchers, even in the humanities, to work collectively.

Although we came across Daston's narrative long after conceptualizing this book, we feel that her story eloquently captures what we would like the book to convey. That is, that interesting interdisciplinary research is not just about bringing together the practices of two or more disciplines to produce a redefinition of topics, questions, and their answers. Its quality also depends on the right attitude of researchers, such as the courage to try something new or an openness to unexpected findings, and on a permissive environment that not only allows for finding inspiration but also allows for doing seemingly mundane work or taking a potentially wrong turn. Daston's story illustrates the lasting effects of inspiring examples on a scholarly trajectory of exhilarating interdisciplinary collaboration and writing.

# The Kid That Transformed the Block

In academic research and teaching, interdisciplinarity is no longer the new kid on the block—it is widely preached and practiced. Definitions of multidisciplinarity, interdisciplinarity, and transdisciplinarity abound (Evers et al., 2015). We know of great introductions to the logics behind and modes and implications of interdisciplinarity (Barry et al., 2008; Barry & Born, 2013), as well as overviews of the history and theory of the concept's use and knowledge ecology (Moran, 2010; Frodeman, 2017). Other publications critically assess interdisciplinarity as a form of scientific imperialism (Mäki et al., 2019), draw lessons from failures in interdisciplinary research (Fam & O'Rourke, 2020), delve into its epistemic pitfalls (Hvidtfeldt, 2018), or focus on the reasons for and problems of a particular interdisciplinary matrix, such as the intersection of social science and neuroscience (Callard & Fitzgerald, 2015).

Recently published handbooks instruct readers in depth on how to develop interdisciplinary research—their chapters interspersed with brief examples in tables and boxes (Repko, 2012)—or discuss the ins and outs of interdisciplinary careers (Lyall, 2019). Highly reflexive is a recent book by Celia Lury et al. (2020) on the process of pursuing interdisciplinary work, which distinguishes, for instance, between the practices of making and assembling, and valuing and validating in interdisciplinarity. We will return to some of the literature on interdisciplinarity when we discuss the most significant themes that run through the three parts of this book. But first, it is necessary to elaborate on what we would like to achieve with this publication.

# Interdisciplinarity in the Scholarly Life Cycle

This collection of essays aims to show how interdisciplinary research develops *over time* in the lives of scholars, not in a single project, but as an attitude that gradually trickles down or spirals up during our practice as researchers. It documents how interdisciplinary work has inspired shifts in how we read, value concepts, critically combine methods, cope with knowledge hierarchies, adopt writing styles, and collaborate. We do so by starting from examples, hence the book's subtitle. These examples have the humanities and social sciences at their core, but they also showcase connections and collaborations with the arts, the medical field, the natural sciences, and computer science. The authors show how they began, attempted to open up, dealt with inconsistences, had to adapt, and learned so much that their approach to research was altered with lasting effects. They also show how they could have developed their work even further had they known what they do now. Our book is thus not *about* interdisciplinarity, but shows *how* it can be practiced by offering a behind-the-scenes approach.

Our show-rather-than-tell approach implies that it is not our goal to propose our own definitions of inter-, multi-, and transdisciplinarity. Most of our authors implicitly or explicitly follow the "modes of interdisciplinarity" conceived by Barry et al. (2008). In this typology, research in the "subordination-service mode" combines methods and concepts from two or more disciplines without aiming to achieve symmetry in the original disciplines' assessment of what is good research. In those cases, the expertise from some disciplines functions as heuristic tool in relation to others. For example, dendrology has this kind of subordinate service relationship to archeology. Most scholars consider this type of research to be multidisciplinary (Evers et al., 2015, p. 13). In contrast, research in what Barry and his colleagues label the "integrative-synthesis mode" *does* judge the integrative work according to the criteria of the disciplines that feed into it. Research in the "agonistic-antagonistic mode," finally, aims at contesting or transcending "the epistemological and/or ontological assumptions of specific historical disciplines" (Barry & Born, 2013, p. 12), thus potentially shattering the ground on which these disciplines stand.

In a typology suggested in the 1990s, science and technology studies (STS) scholar Geoffrey Bowden considered such forms of critical reflexivity "transdisciplinary" in character, often permeating a wide range of domains within the humanities and social sciences. His example was postmodernism and its interest in reflecting on assumptions about knowledge production through the use of new literary forms (1995, p. 69). Other authors, however, reserve the notion of transdisciplinarity for collaborations between academics and nonacademics such as artists (Fam & O'Rourke, 2020, p. 2). When seriously considered, such collaborations may, in fact, imply the epistemological and ontological shifts Barry and his colleagues (2008) had in mind when identifying the agonistic-antagonistic mode.

We hope that opening up our experiences with interdisciplinarity by providing extensive examples and their lasting effects on research is helpful not only for other researchers but also for teaching the next generation of scholars. We have noticed that students seem to understand the concept of interdisciplinarity quite well in theory. In one of our own courses on research skills, for instance, we play a game in which students connect research questions to these questions' disciplinary backgrounds. Most students do well in this exercise, often insightfully reflecting on their choices. Yet they tend to forget these insights as soon as they have to formulate their own research questions. In addition to referring students to some of the how-to-do-interdisciplinary research manuals already mentioned, we hope that educators can refer them to one of the chapters in this book as well. By offering examples across a wide range of topics in the humanities and social sciences, and with a clear indication of the disciplines combined in most of the chapter titles, students may focus on the chapters that best match their own research undertakings. In this way, the research examples offer entries into specific ways of conducting interdisciplinary research that may inspire students and support their own scholarly work. This could then be the start of the student's own scholarly life cycle, within or beyond academia.

All the authors included in this collection are affiliated or collaborate with scholars in the Faculty of Arts and Social Sciences (FASoS) at Maastricht University. This faculty then functions as an institutional hub for the book's examples. Established in the 1990s, FASoS has pursued its research and teaching by drawing on an organizational matrix intended to foster interdisciplinarity. Interdisciplinary research and teaching programs, rather than individual departments, form the faculty's core. Although most departments still have disciplinary names, they often host scholars from a wide range of disciplines. The faculty's spatial organization is also significant: rather than clustered into departments, research groups, or teaching programs, offices are randomly assigned. The rationale behind this was a frequently employed argument to justify interdisciplinarity—that is, that problems in modern society are so complex that they can only be properly dealt with by traversing the borders dividing traditional disciplines. The need for border crossing should then also be visible in the faculty's structure and spatial organization.

This specific goal and organizational structure have attracted academic staff and students from all over the world, making FASoS the most international humanities and social sciences faculty in the Netherlands. Newcomers are often pleasantly surprised to find that scholars really do work in an interdisciplinary manner here and that this approach is supported and assessed accordingly. What is key to this spirit is the way that FASoS manages its teaching. All courses are designed by small groups of teachers with backgrounds in different disciplines. As their courses often remain on offer—be it in modified versions—for many years, the teachers engage in prolonged interdisciplinary collaboration. In this way, they mutually and positively "infect" each other in ways that go beyond the more conventional project-based interdisciplinarity, which is, as Fam and O'Rourke (2020) have explained, more prone to failure.

This does not imply that FASoS teachers, researchers, and their collaborators never encounter the disadvantages of pursuing interdisciplinary careers. Prior to requesting abstracts for this publication, we spoke extensively with all the scholars potentially interested in contributing essays about their experiences with interdisciplinarity. Some of them noted that colleagues had warned them to avoid solely publishing in interdisciplinary venues because that might thwart their future career possibilities in monodisciplinary institutes. Nonetheless, most colleagues gradually adapted and began to embody a scholarly life marked by interdisciplinary teaching and research. As Felicity Callard and Des Fitzgerald underline, the career risks of interdisciplinarity "aren't what they used to be," and staying where you are when "the plate tectonics of the human sciences are shifting" is risky as well (2015, pp. 12–13).

# Finding Your Way in This Book

What becoming an interdisciplinarian means and implies in practice is the subject of this book, which is not unlike the ideal of recovering "detailed actions and reasoning" in the practice-oriented approach that Guelfo Carbone et al. (2019, p. 5) adhered to when discussing experiments of art and science collaborations. By illustrating how our working-in-interdisciplinary-ways *developed over time,* we aim to achieve three additional goals. The first is to inspire both students—at the undergraduate, graduate, and postgraduate levels—and colleagues to perform interdisciplinary research without underestimating the potential challenges of such work. No matter how much we love what we do, we should articulate, rather than suppress, the complexities we attempt to resolve, whether we are successful or not. The second is to show under what conditions interdisciplinarity is able to thrive in academic settings; we have already partially indicated above what supports such sustainability. The third is to illustrate what insights may result from performing interdisciplinarity, be it anthropology in history, philosophy in innovation studies, history in development studies, or arts in sciences, to mention just a few possibilities. Even though we fully acknowledge that such disciplinary labels are subject to historical change, we have used them in our chapter titles to help readers find work intervening in the disciplinary intersections of their interest.

We have divided the book into three sections. The first section, entitled "Moving Concepts: What Theory Can Do," includes contributions that illustrate how and for what reasons theoretical concepts can be made to traverse disciplinary boundaries, from philosophy to science and technology studies, sociology and law to history, sociology and the history of ideas to European studies, and history to development studies. In so doing, these concepts help generate insights that would not have emerged within the confines of a specific discipline.

In the first two chapters, by Harro van Lente and Jo Wachelder, respectively, the authors look back at transformative moments of the early stages of their careers, when the discovery of particular frameworks outside of their disciplinary homes proved fundamental to their research.

Van Lente shows how the philosopher José Ortega y Gasset's notion of "life as drama" altered his research on expectations in technological innovation. This philosophical detour afforded him a better understanding of the direction of innovation, paving the way for what is now called the "sociology of expectations." Wachelder's chapter demonstrates how Niklas Luhmann's sociological theories, especially those pertaining to self-organizing social systems, helped refine his examination of educational debates about the modernization of higher education in the nineteenth-century Netherlands. The application of a discursive analytical approach based on Luhmann's theory allowed him to reconsider the presumed influence of Wilhelm von Humboldt's educational philosophy in the Netherlands.

Karin van Leeuwen's contribution also relates to a historical interest, namely the author's study of constitutional reform in the postwar Netherlands. Van Leeuwen describes how she drew on critical legal studies and Bourdieusian legal sociology to arrive at a more comprehensive narrative of constitutional reform. She argues that law follows but also shapes political negotiation. Writing about the political history of constitutional reform therefore requires more than just mapping out content-related details of the development of the law or reducing debates to clashes between political parties. Ferenc Laczó's chapter shows how a historical-critical approach toward the idea of Eastern Europe helps frame and understand the challenges facing Europe today, while also prompting European studies to rethink the hierarchies of power and knowledge embedded in the field. Employing an idea-historical perspective, Laczó builds on insights from history, geography, linguistics, religious studies, economics, and politics. For Van Leeuwen and Laczó, both trained as historians, translating and implementing theories and concepts from other fields have given new directions to their scholarship and professional trajectories.

In his chapter, Paul Stephenson applies sociologist Marcel Mauss's theory of gift and reciprocity to an analysis of a public policy experiment in France: the establishment of the annual Solidarity Day in the aftermath of the 2003 heat wave that killed numerous older people. Stephenson recounts how early in his career, before he was fully trained in political science, he was excited to discover Mauss's theory and its application to social anthropology and aimed to make it relevant to the study of public policy and crisis management in a modern Western European state. Finally, in her chapter Elsje Fourie shows how the historical concept of low modernism enabled her to develop new perspectives on recent Japanese-Ethiopian development cooperation—specifically regarding attempts by Japanese aid agencies to transfer Japanese management techniques (kaizen) to Ethiopian factory floors. This historical exploration prompted Fourie to reconsider the identity and purpose of development studies as an interdisciplinary field.

The second section of this book, "Refolding Methods: How Twists Require Tweaks," contains essays about how the discovery of methods from fields outside of one's comfort zone prompted these authors to work with different ways of collecting, eliciting, and analyzing data. Interestingly, ethnographic approaches are crucial to the interdisciplinary development of three humanities scholars. This section also includes a chapter that reveals how the development of an interdisciplinary conceptual framework can result in methodological innovation.

The chapters by Aagje Swinnen and Emilie Sitzia demonstrate how two literary scholars with a focus on narrative were inspired by the ethnographic approaches that characterized the fields of gerontology and museum studies. Swinnen describes how she moved away from close readings of representations of aging to experimenting with a reading and writing club for people over sixty that discussed diverse novels about aging. This approach allowed her to both gain insights into older people's attitudes toward and experiences of aging and deepen her understanding of the cultural work performed by novels that address aging. Sitzia, in turn, details the amalgam of methodological approaches from different disciplines that the author developed to study multimodal storytelling in the exhibition "Connectivities" at Mucem (The Museum of European and Mediterranean Civilizations) in Marseille. The piece demonstrates the necessity of such a mix to fully understand the dynamics between the creation, materialization, and reception of the exhibition's narrative, which museums can also learn from in order to optimize the visitor experience.

In Karin Bijsterveld's chapter on her early research of postwar elderly Dutch homes, she demonstrates how she was inspired by the ethnographic work of the Indian anthropologist Sanjib Datta Chowdhury in the same context. This prompted her to more closely study the architectural plans and photographs that were part of the policy documents under study, which eventually transformed her own research premises. The final chapter in this section, by Kathleen Gregory, Paul Groth, Andrea Scharnhorst, and Sally Wyatt, presents the project Re-SEARCH, which brought commercial and academic partners together to investigate and develop search solutions for research data. The authors explain how they consolidated an innovative conceptual framework by synthesizing different notions of users to facilitate interdisciplinary collaboration between STS researchers and computer scientists, and between designers of data search systems and their users.

The third section of the volume, entitled "Cascading Collaborations: With Artists, Style, and Skill," focuses on the so-called collaborative turn in humanities and social sciences scholarship. This turn is indicative of the collaborations across disciplines that interdisciplinary scholarship requires. The chapters in this section examine the different challenges and rewards of diverse strands of collaborative work. They also discuss how earlier experiences with collaboration inform later ones, hence the notion of "cascading."

The first three chapters look at what collaborations between partners from the arts, humanities, and sciences entail. The chapter by Flora Lysen starts from the author's observation of the collaboration between the neuroscientist Alexander Sack and the artist Antye Guenther in the Maastricht Brain Stimulation and Cognition Lab. She explains the tacit affective dispositions characteristic of what she calls the "imaginary of the inter," a shared sense of collaboration that enables diverging objectives and expectations to productively coexist. Patricia de Vries examines "Reprodutopia," a 2019 exhibition with a prototype of an artificial womb and the scenarios it engenders for the future of reproduction, developed by speculative designers from Next Nature Network in collaboration with scholars from the Máxima Medical Center and the Eindhoven University of Technology. De Vries demonstrates the necessary interdisciplinary work to reveal how imminent technologies build on specific sociotechnical and medical histories as well as its consequences and limitations. In the chapter that follows, Peter F. Peters, Ties van der

Werff, Imogen Eve, and Jos Roeden reflect on a collaboration between the South Netherlands Philharmonic, the Conservatorium Maastricht (a higher arts education institute), and Maastricht University that aimed to show how symphonic orchestras can shape new futures by innovating their music practices. Inspired by the work of Richard Sennett, the authors address the role of dialogic versus dialectical conversations in collaborative research on new concert formats and audience participation.

Jessica Mesman's contribution also focuses on practice optimization, albeit in two healthcare settings: the emergency department of an Australian hospital and a Dutch maternity ward. Mesman demonstrates how the implementation of video-reflexive ethnography (VRE) in the study of daily healthcare routines transforms practitioners into co-researchers. She argues that VRE, as a tool for exnovation—bringing out what the practitioners already know—rather than innovation, both articulates and overcomes disciplinary and paradigmatic differences. To exemplify these processes, she discusses issues of professional credibility and reputation that are at stake in collaborative work.

The final chapter of the publication departs from the format of other chapters by taking the form of a conversational piece. Valentina Mazzucato, Bilisuma Dito, and Karlijn Haagsman have pursued a longstanding interdisciplinary collaboration on the topic of how transnational immigrants "do family" when their relatives are separated by great geographical distances. They use the metaphor of "doing family" to reflect on the practice of teamwork along the themes of open communication, trust, and friction. This chapter elucidates the emotional work and soft skills regarding attitudes and predispositions toward people and disciplines that collaboration inherently demands.

Traversing all three sections are affinities between the chapters in relation to the substance of the research examples discussed. Such affinities concerning the topics presented are aging (Bijsterveld and Swinnen), the visual arts and design (Lysen and De Vries), issues of development and migration (Fourie and Mazzucato c.s.), innovation and exnovation (Van Lente, Gregory et al., and Mesman), nation-state politics and ideas (Stephenson and Laczó), the logic of law (Van Leeuwen and Wachelder), and institutes of heritage conservation (Peters et al. and Sitzia).

Finally, all three sections include a few illustrations crafted by chapter authors who dared to go beyond their writing skills and tried out drawing or making a collage to capture what their interdisciplinary work was about. They did so in the inspiring companionship and with expert advice of art teacher Marte Hameleers and photography teacher Maaike Faas-Schauer. The creative exercise helped the authors to capture a key message of their chapter while enacting transdisciplinarity, which in turn often re-informed their writing. The same happened when we asked Eric Bleize, who is both a scheduler at Maastricht University and a photographer, whether he would allow us to use one of his art photos for the book's cover. With his permission, for which we are grateful, we chose one of his multi-exposure photos. It shows both the gate to an academic building and the movements of those entering and passing it. To us, the port stands for the entrance into scholarly life at large, while the movements around it signify the many shifts that interdisciplinarity brings to academic research and learning.

## Recurring Themes

We would like to conclude our introduction by elaborating upon three themes that recur throughout the sections: redefinition as a key goal or result of interdisciplinary research, curated curiosity as an important tool for getting there, and sustained collaboration as its necessary condition. These themes support some of the claims in the secondary literature mentioned in our opening paragraph but also occasionally depart from arguments made in this literature. In making such comparisons, we do not claim to exhaustively cover the by now extensive body of secondary literature on interdisciplinarity. However, we do seek to highlight the key characteristics of how contributing authors have conducted interdisciplinary work against the background of how other scholarly literature represents such work.

First, nearly all authors are out for or end up with a *redefinition* of a topic, a key concept, an established hierarchy, a method, or even an entire field's objective—by engaging in some sort of interdisciplinary integration. Mesman's embodiment of STS in medicine through VRE leads to studying exnovation rather than innovation. Laczó re-centers Eastern Europe while also broadening what belongs to the study of politics. Fourie's use of the low modernization concept reveals development studies' blind eye for the Western roots of the Japanese imposition of kaizen on Ethiopian shop floors and, thus, the definition of what counts as East and West, North and South. Implicitly or explicitly, such redefinitions function as the main marker of the authors' interdisciplinary success, while their ability to explicate the integration work behind it serves as its secondary marker.

In this way, contributors try to avoid the "interdisciplinary Halloween" outlined by Jonathan Sterne in a 2007 blog entry that critically responded to interdisciplinarity as a management ideal. What Sterne argued against was a form of quasi-interdisciplinarity, in which the intellectual reasons for integration are lost or in which scholars just import the work of other disciplines into their own without acknowledging these other disciplines' traditions. What was not that relevant for our authors, however, was the distance in terms of approach between the disciplines involved. While Rolf Hvidtfeldt finds it "unimpressive" to talk about interdisciplinarity when two approaches share too many "paradigmatic examples of good practice" (p. 22), our authors consider interdisciplinary integration *within* the humanities and social sciences as no less adventurous than integration between, for instance, the humanities and the sciences. This is even true for work in which they combine several interpretative traditions. Their sensitivity for differences in "narrow" interdisciplinarity (Klein, 1990, as cited in Hvidtfeldt, 2018, p. 22) is nevertheless instructive in two ways. It helps to articulate the ways in which seemingly similar approaches still differ in their disciplinary take on the subjects under examination, and it is educational in unraveling the type of issues at stake when doing interdisciplinary research.

We need to add two caveats though. One is that in our discussion with the authors prior to the chapter writing, some argued that they never considered themselves monodisciplinary scholars to begin with but as "interdisciplinarians" right from the start. The other is that several of the authors do not see achieving interdisciplinary integration and strengthening their disciplinary identity as mutually exclusive. In their view, the two go hand in hand. For example, taking up the work of a philosopher only intensified Van Lente's devotion to innovation sciences. By acknowledging future-oriented imaginaries, Van Lente succeeded in altering his field in such a way that he actually felt *more* at home there. Bijsterveld's engagement with an anthropological interpretation of architectural design made her attentive to what architectural plans perform. In hindsight, it clarified what "acts of notation" do, as one of the skills that may laterally move between disciplines and thus contribute to interdisciplinarity (Wedell, 2020, p. 117). It also allowed her, however, to reconfirm her practice as a historian of tracing phenomena over time, now with an intensified attentiveness to changing conceptions of such phenomena. The next interactions would never be the same, but future interdisciplinary partners would definitely still be engaging with a historian. These are hopefully examples of what Thomas Osborne has called "trespassing" on "one's own" or "interdisciplinarity in one person" (2013, p. 88).

But how do we keep from becoming overwhelmed by the potentially dizzying exercise of an interdisciplinary practice? Recent literature that guides students through interdisciplinary research suggests a step-by-step approach to identifying an object of investigation, formulating interdisciplinary research questions, justifying their usefulness, and selecting relevant disciplines for one's literature search (Repko, 2012, pp. xxviii–xxxi, pp. 84–89). No matter how useful such advice is, especially when cast in the deep recognition of the iterative character and reflexivity of research, several of our authors underline the importance of an interdisciplinary culture that offers curatorial guidance as well as the freedom to explore literature beyond an already established canon. A theoretical physicist by training, Wachelder was new to the world of history, philosophy, and sociology when he initiated his research. The welcoming attitude of his peers and their generosity in sharing their expertise inspired him to write his system-theory informed history of the university.

Such experiences should not be read as a suggestion to simply skip a solid literature search as a formative phase of the research process—if only to prevent reinventing the wheel in a particular domain of scholarship. However, the idea that the search should *begin* with a phenomenon tends to neglect three important points. First, genuinely original questions usually intervene in the boundaries of what the phenomenon "is," in what is supposed to be inside and outside the object of study, as Daston's reading of Hacking's work already illustrated. Second, compelling questions often result from literacy in pockets of loosely connected literature that require years to become familiar with. And finally, it is often the perceived ethos of the academic curators of those literatures that make both junior and senior scholars take the offered interdisciplinary threads seriously or not. While Repko highlights that interdisciplinarity requires a deep understanding of the relevant disciplines' epistemologies, theories, concepts, and histories, acquiring familiarity with the interstices of the fields in question takes much more than just one project.

Collectively stimulating a *curated curiosity* for continuous reading—our second recurrent theme—is, therefore, a highly relevant tool for doing solid interdisciplinary research. It is an inconspicuous dimension of interdisciplinary work, but the stories told in our book flag its relevance. It is no coincidence that several of the authors embarked on their interdisciplinary journey with a classic and, therefore, ubiquitous study from a field they had just discovered—sometimes only later recognizing the work's defining role in that field. To scholars from these other fields, this kind of first encounter with a canon they are so familiar with may seem unoriginal. However, observing such excitement may also entail something akin to mild jealousy—just as one may envy a novice reader of Leo Tolstoy's work. More importantly, a proper contextualization and compelling re-embedding of canonical works may offer novel insights, such as Stephenson's uptake of Mauss's *The Gift* (1954) for policy studies or Fourie's enthusiasm about Jess Gilbert's notion of low modernism (2003) for development studies show.

Finding such resources may often result from the casual browsing through library stacks that Van Lente recounts in his chapter. It can also result from the "library brachiation" that sociologist Andrew Abbott describes in his book *Digital Paper* (2014, p. 22). There, he recalls his early visits to libraries, where the call numbers assigned by librarians, as old-school curators, indicated which books were stacked together. Abbott deliberately departed from such curatorship by perusing the footnotes of crucial books for references to other relevant publications, checking their (often unexpected) call numbers, and then adding these to his search—hence the idea of brachiating. As he notes, brachiating requires other techniques in the digital era, which only underlines the need for an institutional culture that fosters interdisciplinary curating while stimulating an open attitude about what might turn out to be important for interpreting one's primary materials.

This is all the more important for the humanities and qualitative social sciences—most notably cultural studies and literary scholarship—which often make use of what Hermann von Helmholtz termed "aesthetic induction" in 1862. He defined aesthetic induction as the opposite of logical induction and a manner of argumentation that embodied the specific usefulness of the humanities. While logical induction is a systematic process drawing on assumptions and rules, aesthetic induction leaves such rules behind, pioneering beyond them. In the words of philosopher Rein de Wilde, aesthetic induction is all about association, about "ideas that occur to you as in 'Ah, this reminds me of …'" (De Wilde, 2012, p. 288). This way of working is visible in De Vries's contribution, which describes an arts-science collaboration that led to the exhibition design of an experimental artificial womb. To elucidate the performative effects of such an exhibition, De Vries shows how its representation of the artificial womb drew upon age-old imaginings of the womb, while also broadening conceptions about parent-child relationships, which seemed inconceivable until this event. The extent to which such an exercise results in a convincing or original argument depends on the richness of the associations and alignments that authors convey as well as their erudition. This is why interdisciplinarity cannot be effectively pursued without consistently reading a wide range of sources and developing ways to recall or retrieve all that information—the latter perhaps the least transparent and commented upon aspect of a long academic trajectory.

Third, the necessity of *sustained collaboration* is a message emanating from the pages of many a chapter. This holds true both for the essays that illustrate interdisciplinarity in one person, or as Hvidtfeldt has it, the "polymath-mode" of interdisciplinarity and for the "entirely social modes" (2015, p. 24). Even where individuals bring fields together, their narratives show how colleagues informed them. Interdisciplinarity that draws on collaboration between individuals from different disciplines and backgrounds, however, requires even stronger versions of prolonged institutional support, as well as intellectual investment, interpersonal trust, and a sense of equality among the participants.

In practice, however, interdisciplinarity not always sides with equality of some form. Pleas for interdisciplinarity may actually accompany disguised forms of scholarly imperialism, as Uskali Mäki et al. (2019) have argued. Their case in point is the rise of neuroscience in the social sciences and humanities that have quickly seemed to make it the standard for good research. The application of artificial intelligence to a wide range of subjects once considered to be solely within the purview of the humanities—such as identifying the authenticity of art works (Berezhnoy et al., 2007)—might serve as another example. According to Clarke and Walsh (2009), such forms of interdisciplinarity only deserve to be called imperialistic if the result is that the methods once predominant in the humanities or social sciences are considered invalid. Anything else is just innovation and scientific progress. We would like to add, however, that although one should not use the notion of imperialism too loosely, one must remain alert to what happens during interdisciplinary grant evaluation panels, for example. The validation of the latest and most novel techniques is characteristic of the sciences, a tendency that often clashes with the humanities' valorization of the scholarly past.

Quite a few of the authors encountered issues of epistemic authority when practicing interdisciplinarity—a possibility that Fam and O'Rourke (2020) have warned about. Although Mesman has a medical background in nursing and is often invited by medical experts and health scientists to do VRE in hospital settings, she shows how even something as seemingly mundane as writing a literature review by (not) appropriating the style of the health scientists with whom she collaborates may potentially negate the ethnographer's authority and legitimacy. Mazzucato, Dito, and Haagsman demonstrate that even teams that fully embrace interdisciplinary collaboration as a credo may have to cope with the effect of established hierarchies and encounter mutual mistrust if, in the intensity of creating, for instance, questionnaires together, views are dismissed too readily as irrelevant. As Regina Bendix, Kilian Bizer, and Dorothy Noyes have noted, this may not be surprising in an academic context, as academia is *also* about suspicion, "suspicion of received knowledge, suspicion of other colleagues' arguments, suspicion of oneself and one's own representations" (2017, p. 57).

What is certain is that interdisciplinarity needs time to develop. When Peters and his team designed a theater setup that reflected a collaboration between humanities' scholars, musicians, and audiences, the routines of technicians and musicians resulted in unanticipated pushbacks. While these scholars have usually managed to resolve most issues along the way, Wachelder very openly recounts how an article he submitted to a history journal was rejected by the editors because they felt that the sources he used did not count as archival materials—something he had to take into account in future submissions.

A prolonged immersion in disciplines beyond one's home discipline also invites new interdisciplinary initiatives. Several authors did not so much suggest mixed methods in response to issues that transgress the boundaries of traditional disciplines but, rather, to their research subjects' *own* reflexive tendency to broaden their scope. Sitzia, for instance, notes how museums today reflect on globalization or history making itself rather than merely represent their heritage within a particular epoch. Analyzing how they accomplish this requires a wider palette of methods. Swinnen identified a social trend—reading groups for older people—that responded to the belief (influenced by literary studies) that representations of diversity in fiction have the performative effect of denaturalizing stereotypes and valuing alternative ways of life. Critically examining this claim required integrating social science and humanities methods.

The recognition of and sophisticated response to differences in writing styles is once again dependent on sustained collaboration, and somewhat underrepresented in the literature on interdisciplinarity. A handbook by Allen F. Repko (2012), for example, does not mention this issue in what is otherwise a very comprehensive introduction. When Bijsterveld entered the field of STS as a historian, she noticed that STS practitioners often used italics to highlight analytical distinctions between concepts. A paper without many italicized words, then, seemed to not conform to the field, although none of Bijsterveld's STS colleagues explicitly instructed her about this. Such conventions could only be learned through practice. The same is true for the essayistic, narrative style of the humanities versus the descriptive, empirical style of the social and natural sciences, as Mesman experienced extensively.

Self-confidence in the offerings as well as writing styles of the humanities and qualitative social sciences is a key virtue of our interdisciplinary encounters. Lysen demonstrated this when thinking about what she, as an interdisciplinarian, could offer in her analysis of a collaboration between an artist and a group of neuroscience scholars. First and foremost, she used her abilities and courage as a scholarly writer to make the moments of discomfort in this collaboration transparent while situating those moments in the wider net of encounters between art and science. Finally, Gregory et al. show how elegantly hammering down the message that the user of information and computer science research data should never be an implied user but a user whose meaning-making processes are key to research data employment can entice the world of science to accept a patchwork of quantitative and qualitative methods.

## Conclusion

Similar to how Gregory and colleagues redefine what it is to examine the user of technology, other authors redefine their object of research and more. If redefinition is what they are after in their interdisciplinary endeavors, many of them see forms of curated curiosity and sustained collaboration as ways to reach that goal. Their rich examples, however, show what this means in practice.

It has long been acknowledged that scholars have limited control over how their academic work is cited and utilized by others, no matter how well their rhetorical skills are developed. This is even truer for interdisciplinary work, as scholars and nonacademics from many other domains and fields may align and connect to it in unexpected ways.

This is exactly how it should be. Most biographies that scholars submit to conferences and publications are rather conventional and formulaic, whereas scholarly life cycles are full of dead ends, surprising turns, and unexpected uptakes. This was also one of Daston's messages when writing about her intellectual past. Careers are usually not as coherent as résumés suggest (2021, p. 80). Collaborations are among the possible contingencies.

With this in mind, the contributors to our book who worked in teams have written prosopographies, or group bios, instead of individual bios of their collaborative histories. Daston added to her remark about résumés that distractions are key to academic work. It is our hope that this book may offer readers a worthy distraction.

# References

Abbott, A. (2014). *Digital paper: A manual for research and writing with library and internet materials.* The University of Chicago Press.

Barry, A., & Born, G. (Eds.). (2013). *Interdisciplinarity: Reconfigurations of the social and natural sciences.* Routledge.

Barry, A., Born, G., & Weszkalnys, G. (2008). Logics of interdisciplinarity. *Economy and Society, 37*(1), 20–49.

Bendix, R., Bizer, K., & Noyes, D. (Eds.). (2017). *Sustaining interdisciplinary collaboration: A guide for the academy.* University of Illinois Press.

Berezhnoy, I. J., Postma, E. O., & van den Herik, H. J. (2007). Computer analysis of Van Gogh's complementary colours. *Pattern Recognition Letters, 28,* 703–709.

Bowden, G. (1995). Coming of age in STS. In S. J. Jasanoff, G. E. Markle, J. C. Petersen, & T. Pinch (Eds.), *Handbook of science and technology studies* (pp. 64–79). Sage Publications.

Callard, F., & Fitzgerald, D. (2015). *Rethinking interdisciplinarity across the social sciences and neurosciences.* Palgrave Macmillan.

Carbone, G., Gisler, P., & Sormani, P. (2019). Introduction: Experimenting with "art/science"? In P. Sormani, G. Carbone, & P. Gisler (Eds.), *Practicing art/science: Experiments in an emerging field* (pp. 1–18). Routledge.

Clarke, S., & Walsh, A. (2009). Scientific imperialism and the proper relations. *International Studies in the Philosophy of Science, 23*(2), 195–207.

Daston, L. (2021). Families, genealogieën en archieven. *Nexus, n.v.* (88), 67–80.

de Wilde, R. (2012). Wiel Kusters als uitvinder. In J. de Roder (Ed.), *Ik woon in duizend kamers tegelijk: Opstellen voor en over Wiel Kusters* (pp. 287–289). Vantilt.

Evers, A., Jensen, L., & Paul, H. (2015). *Grensverleggend: Kansen en belemmeringen voor interdisciplinair onderzoek.* Koninklijke Nederlandse Akademie van Wetenschappen/De Jonge Akademie.

Fam, D., & O'Rourke, M. (Eds.). (2020). *Interdisciplinary and transdisciplinary failures: Lessons learned from cautionary tales.* Routledge.

Frodeman, R. (Ed.). (2017). *Oxford handbook of interdisciplinarity*. Oxford University Press. (Original work published 2010).

Hvidtfeldt, R. (2018). *The structure of interdisciplinary research*. Palgrave Macmillan.

Lury, C., Fensham, R., Heller-Nicholas, A., Lammes, S., Last, S., Michael, M., & Uprichard, E. (Eds.). (2020). *Routledge handbook of interdisciplinary research methods*. Routledge.

Lyall, C. (2019). *Being an interdisciplinary academic: How institutions shape university careers*. Palgrave Macmillan.

Mäki, U., Walsh, A., & Fernández Pinto, M. (Eds.). (2019). *Scientific imperialism: Exploring the boundaries of interdisciplinarity*. Routledge.

Moran, J. (2010). *Interdisciplinarity*. Routledge.

Osborn, T. (2013). Inter that discipline! In A. Barry & G. Born (Eds.), *Interdisciplinarity: Reconfigurations of the social and natural sciences* (pp. 82–98). Routledge.

Repko, A. F. (2012). *Interdisciplinary research: Process and theory*. Sage.

Sterne, J. (2007). *On interdisciplinarity*. https://superbon.net/2008/11/15/on-interdisciplinarity/

Wedell, M. (2020). Notating. In C. Lury, R. Fensham, A. Heller-Nicholas, S. Lammes, S. Last, M. Michael, & E. Uprichard (Eds.), *Routledge handbook of interdisciplinary research methods* (pp. 116–121). Routledge.

# Part I

## Moving Concepts: What Theory Can Do

# Reversing the Gaze on Expectations in Technology: The Philosopher José Ortega y Gasset and Innovation Studies

Harro van Lente

## Introduction

How do we explain the direction of innovation? In the early 1990s, this was one of the key questions in the field of innovation studies, which I also pursued in my own research at the time. It was a novel question since traditionally, innovation studies had endeavored to explain the speed of innovation. In my research, I followed the role of expectations in innovation in order to contribute to this and other new questions in the field of innovation studies. In this chapter, I will elaborate on how a casual reading of the work of the philosopher José Ortega y Gasset (1883–1955) spawned a breakthrough in my thinking on the subject matter. While his ideas on life as a drama first appeared as an interesting but not so relevant detour adjacent to the core of my research, it forged a crucial reversal in my research on expectations in technology. This reversal, in turn, helped

H. van Lente (✉)
Department of Society Studies, Faculty of Arts & Social Sciences,
Maastricht University, Maastricht, The Netherlands
e-mail: h.vanlente@maastrichtuniversity.nl

K. Bijsterveld, A. Swinnen (eds.), *Interdisciplinarity in the Scholarly Life Cycle*,
https://doi.org/10.1007/978-3-031-11108-2_2

me to formulate a "sociology of expectations" as a novel approach to innovation studies. My primary claim, thus, is that interdisciplinary detours may prove decisive.

In this chapter, I will first sketch the agenda of innovation studies in the early 1990s and introduce the question of direction of innovation. Then I will turn to my reading of Ortega y Gasset and its conceptual implications for my research on expectations. I will also reflect on the conditions for this fortunate philosophical detour and what it says about the intellectual adventure of interdisciplinarity.

Innovation studies is a scholarly field investigating how technological innovations are developed and used (Fagerberg & Verspagen, 2009). There is a strong emphasis on what happens within firms that depend on innovation for their ability to compete. Many firms have large departments for Research and Development (R&D) with budgets for this so-called private research that are comparable to those of public research at universities. Industrial firms are interested in the success factors of innovation projects, as their success is notoriously difficult to predict and manage. Likewise, governments are interested in the general conditions that help innovation, hoping to promote a competitive national economy. Within the field of innovation studies, there is also interest in the consequences of technologies and the societal meaning of innovation in economic, social, and cultural terms.

## Innovation Studies and the Problem of Direction

Until the 1980s, the central question in innovation studies was to explain the pace of innovation: how to foster innovation processes in firms and gauge how quickly innovations would spread in society. Since the 1960s, the standard approach has been to follow the example set by Everett Rogers (2003). In *Diffusion of Innovations*, originally published in 1962, Rogers presented a model of the spread of innovations in society, which entailed a slow uptake in the beginning followed by a steady increase and a slowing down once "saturation" was reached. In many cases, the introduction of new technologies indeed follows this pattern. Television sets or mobile phones, for instance, were initially rare and only used by "early adopters." Gradually, more and more customers followed, with numbers increasing very quickly but then later slowing down. Eventually, when almost every household had a television set or several mobile phones, one could speak of saturation. In Roger's model, the spread of many new technologies in society is captured in an S curve, mathematically expressed with a logistic function. His model of the diffusion of technologies was inspired by studies on mass communication about the spreading of influential ideas. In his seminal study, Rogers also distinguished between categories of users, ranging from "early adopters" to "late adopters" or "laggards." The derogatory nature of the latter term suggests a pro-innovation bias.

In the 1980s, the approach to technologies as gradually diffusing into society was challenged by different voices. One strand of new questions came from the philosophy of science and the history of technology. Here, the new impetus was to explain the particular design of a technology. It was no longer seen as sufficient to take the form of a new technology for granted and to only study its adoption by society. The aim was now to open the "black box" of technology, thanks to a new turn in the philosophy of science based on Thomas Kuhn's work (1962) about the seventeenth-century scientific revolution. Kuhn showed that the fate of scientific truths is linked to how communities of scientists think and work. Scientific theories and claims are not in and of themselves true or not, he showed, but can only be understood as a "paradigm," that is, a particular way of seeing reality. When paradigms shift, such as during the scientific revolution when the empirical method became the dominant approach, the direction of science changes as well. It follows, then, that the direction of scientific progress can be explained by looking at scientists' choices. This is as true for "bad" science, such as the infamous genetic theories of Trofim Lysenko underpinning Soviet agricultural policies, as much as it is for "good" science, like James Watson's and Francis Crick's discovery of the double helix structure of DNA.

Some scholars extended these new questions in the philosophy of science to the study of technology and asked: Why do engineers and firms make particular choices? Why are new technologies seen as "working"?

Trevor Pinch and Wiebe Bijker's study (1984) on the development of bicycles at the end of the nineteenth century, for instance, examined how the bicycle acquired its particular shape—as a contraption with two equally sized rubber-tire wheels connected by a chain with pedals between them. Pinch and Bijker studied the diversity of early bicycle designs, which scarcely resembled the modern bicycle. Instead, they discovered an astonishing range of models with wheels that ranged from small to large, with pedals and saddles in different places. These were contenders with their own strong and weak points, at least in the eyes of the various beholders, which Pinch and Bijker termed "interpretative flexibility." This fluid situation continued for several decades, until eventually the diversity of models was replaced by the one we today associate with the bicycle. The direction of innovation was thus explained as a societal struggle between meanings attributed to particular versions of a technology.

Another novel strand in the field of innovation studies was evolutionary economics, with Richard Nelson, Sidney Winter (1977), and Giovanni Dosi et al. (1988) as its key figures. While mainstream economics tends to see innovation as an external, exogenous factor, this new strand of research introduced the idea of interpreting innovation as an economic phenomenon. Nelson and Winter have argued that innovation is one of the ways that firms compete, though it does not always guarantee success. Innovation, they emphasized, is not an outcome of calculable optimization but an inherently uncertain process. Nelson and Winter proposed an evolutionary model of variation and selection: firms launch various designs (variations) that compete in the market (selection). Unlike the biological process, however, variations are not random and selection is not blind. Firms do not innovate randomly but structure their search processes through routines and heuristics in their production of variations; likewise, markets are influenced by marketing efforts and lobbying strategies. The direction of innovation, thus, can be explained by carefully following the routines, heuristics, and attempts to shape markets.

These new strands in innovation studies, in turn, prompted a new question: How do we account for the direction of innovation? These were questions about why scientists, engineers, and firms make particular choices. Sociological studies explained how new technologies take shape through a battle of meanings. Evolutionary economics pointed to the role of routines within organizations and the importance of heuristics. The common idea was that under conditions of uncertainty, hopes about future success would guide researchers and firms in their search activities.

## Reading Ortega y Gasset

When historians, sociologists, and economists opened the "black box of technology," they began to raise new questions. What choices do engineers and firms make? Which designs promise to be successful? How should the very processes of innovation be understood? This was the departure point of my research in the 1990s; from there, I decided to study the role of expectations in the development of technology. The idea was that engineers and firms would relate their choices to expectations of future success. My research was exploratory, conceptual, and empirical in character. In my conceptual exploration, I had been reading into works of evolutionary economics, science studies, and organization studies. In my empirical explorations, I followed scientists and engineers and attended their exhibitions, conferences, and classroom lectures. I delved into their world and probed their efforts; my former training in physics was helpful for this purpose.

During my research, I took up the habit of a regular Friday afternoon visit to the library to browse and nurture myself with novel, amusing, and intriguing ideas. These exercises were not goal-oriented but motivated by curiosity. I simply allowed myself to follow traces that appeared interesting. The ethnographer Paul Rock (1979) once called this the AHFA method, or Ad Hoc Fumbling Around. In this way, I read about architecture, sports, the history of technology, or cultural criticism—and not necessarily the most recent publications. I also came across the work of José Ortega y Gasset, a Spanish philosopher with a poetic name. I read his essay "Man the Technician" (1961), which first appeared in 1930 as *Meditación de la técnica*.

Ortega y Gasset is probably the most important Spanish philosopher of the twentieth century. He was a public figure who engaged with the pressing question of how Spain should relate to the rest of Europe and to

modernization. For a short time before World War II, he was also a par-
liamentarian. During the dictatorship of Franco, he worked as an exile in
Argentina and returned to Madrid in 1948 to establish the Instituto de
Humanidades, dedicated to the humanities and its societal relevance. He
is now known as a "crisis" philosopher, critical about the reductionist and
rationalist approaches that come with modernity and industrialization. A
core idea of his philosophy is "vital reason," defining humanity not just
by reason but also by vitality. He described the human condition as
marked by the urgent need to jump into life, with an overdose of energy
and desire. He is best known for his work of cultural criticism entitled
*The Revolt of the Masses* (1930), which points to the eroding forces of
industrialization and the rise of a new citizen with no duties, only rights.
The right to see things improve is among them: an entitlement to progress.

His *Meditación de la técnica* inspired me deeply because it criticized
standard assumptions about technology and offered an alternative perspective on what it means to be human. The text begins with a critical
reflection on what it is to "need" something, given that technology is
usually seen as an answer to human necessities. So-called instrumentalist
philosophy has further developed this view by claiming that technology
should be understood as an extension of the human body. The idea is
that, in contrast to animals, the human body is not sufficiently equipped
to survive: humans lack fur, sharp nails, speed, and strength. Thus, technology comes to the rescue. Among the proponents of this view, the
German philosopher Arnold Gehlen has characterized the human being
as a *Mängelwesen* (deficient creature), who in contrast to animals needs
technology to survive. Ortega finds these ideas too simplistic and investi-
gates the condition of need. What does it mean that when you are cold,
you need clothing and that when you are hungry, you need food? It
means that you do not want to die. Without clothing and food you will
die, so these things are necessary. Their necessity, Ortega argues, is thus
conditional: The condition is the will to live. The necessity of needs, he
concludes, is not the necessity of a stone falling downward. To live is the
primary need; all other needs are secondary.

The second step in his investigation is a closer inspection of the need
to live. What kind of need is it? He refers to anthropological studies
showing that since the earliest traces of homo sapiens, both "useful" tools

existed as well as "superfluous" ones, such as jewelry or musical instruments, and stimulants like khat and kola nuts. Likewise, there are many indications that fire was not just used for heat and food preparation but also for intoxication (to get drunk or high). Ortega concludes that the need to live cannot be distinguished from the need to have a good life— one with beauty, enchantment, and purpose. Here he clearly deviates from the well-known hierarchy of needs that Abraham Maslow proposed in the 1950s, in which "higher" needs like self-actualization are only addressed when "lower" needs like food and safety are met. Ortega argues that both high and low needs are necessary to live a good life and that technologies help achieve both. Technology should not be seen as a compensation for the deficient human body nor a strategy to cope with a hostile nature but as a means to have a good life. The "good life" is a primary invention from which all others follow and "technology is the production of the superfluous, today as in the Paleolithic age" (Ortega y Gasset 1930/1962, p. 18).

At this point in the text, Ortega points to a crucial difference between animals and humans. For animals, to live means to survive in nature; for humans, the pressure to survive has been minimized by technology. The time saved is filled with activities that are not dictated by biological needs but self-invented pursuits. For Ortega, being human, technology, and the good life are intimately related. Our task, he argues, is not to survive but to fulfill a program. In order to decide what to do, we cannot refer to natural laws or a fixed repertoire of activities, as animals do. Life is something we have to invent ourselves. Life starts with an invented life:

> This invented life—invented as a novel or a play is invented—man calls "human life," well-being … Have we heard right? Is human life in its most human dimension a work of fiction? Is man a sort of novelist of himself who conceives the fanciful figure of a personage with its unreal occupations and then, for the sake of converting it into reality, does all the things he does—and becomes an engineer? (Ortega y Gasset, 1930/1962, p. 108)

These and other quotes really moved me. The idea of "man a sort of novelist of himself" highlights the importance of drama, an imagined life, a narrative with protagonists, and a plot. This is the reality we live and must relate to: "Body and soul are things, but I am a drama, if anything, an unending struggle to be what I have to be" (p. 113); "To live ... that is to find means and ways for realizing the program we are" (p. 116). And, given the urge to realize the program or drama, we must be persistent and inventive: "Man has to be an engineer, no matter whether he is gifted for it or not" (p. 136). So, while denying that technology is a rational answer to social needs—provided by engineers—Ortega concludes that we are all engineers nonetheless.

## Reversing the Gaze on Expectations

Why was reading Ortega's reflection on technology such a powerful experience? Two reasons stand out. First, his style is surprising and refreshing. The text lacks references, it does not introduce a proper problem definition, it does not detail and justify a method, and it contains sparse empirical data. The text simply departs from an idea—an observation, an intellectual puzzle—and then leads the reader through thoughts and suggestions, rejecting certain ideas while making bold claims along the way. While this style is not uncommon in philosophy, it differs markedly from the articles in *Research Policy* or in *Social Studies of Science*, the leading journals in my field of innovation and technology studies. The direct appeal to contemplate and the very urgency to think were refreshing; it contrasted with the formats in which I was trained to write. My encounter with this very different style encouraged me to be more daring in my own research and writing without abandoning the requirements of my research field. Ortega's methodology inspired me to be faithful to my ideas in the very manner they came to me.

Secondly, reading this text was powerful because the ideas resonated with the twists and argumentation I had noted in my encounters with engineers and scientists. It helped me to make sense of and articulate them. For instance, I had become more and more skeptical of the prevailing idea that the work of engineers and researchers is characterized by identifying problems and, subsequently, solving them. It is a standard idea, also adhered to by engineers and researchers themselves, floating around as a truism in policies, newspapers, and teaching. It occurred to me, however, that engineers and researchers might not follow this logic. In many cases, the urgency to act and the directions of their efforts did not come from a persistent problem but were derived from ongoing competition. Engineers and researchers are remarkably keen to know what the latest trends are and what is happening in the rest of the field. In one of my case studies, I looked at the rise of membrane technology, dedicated to the sophisticated filtration of liquids and gases (Van Lente & Rip, 1998). Typical questions I encountered at conferences and in interviews included: What are Shell's plans? What are Japan's innovation programs? How far have the companies in California advanced? The urgency to engage with membrane technology is typically phrased as the fear of missing boats or trains that cannot be stopped. Evidently, firms, researchers, and even governmental actors cannot afford to lag behind in innovation races. Moreover, such competition is typically framed in terms of promises and expectations: What is the promising direction to engage with; where will the competition be in the next few years? When a direction in membrane technology is seen as promising—and the moves of competitors indicate this—there is pressure to respond and act accordingly. New technologies, I concluded, do not derive from a problem; rather, they begin with a promise.

In principle, my findings aligned with the starting points of evolutionary economics and could be phrased in those terms. Indeed, researchers and firms must operate under conditions of uncertainty and—yes—heuristics guide them in their searches. The basic argumentation is that when the information to decide is incomplete, one cannot optimize the decision but must come to a satisfying solution instead, as Herbert Simon (1957) has argued. Also, in my research on expectations, I found that the information of engineers and firms was incomplete (for all kinds of obvious reasons) and that, in their decision-making, expectations helped to fill these gaps. My reading of Ortega y Gasset, however, reversed that perspective. Now, I was able to formulate that firms and engineers do not start with gaps in information to be filled with expectations. Instead, they begin with an imaginary future world in which, say, membrane technology exists, which is taken seriously by firms and countries worldwide—and from this, they decide how this should influence their actions. The rhetorical entity of a promising new field of membrane technology is then filled in with actual work (Van Lente & Rip, 1998). Through the promise, funding becomes available, colleagues become interested, and competitors feel the pressure to take it seriously. Increasingly, membrane technology becomes more real, exerting more pressure on engineers, firms, or governments (Van Lente, 2000). Efforts by others are then considered proof that membrane technology *really* exists and demands more attention and greater efforts. Even my own research into the rhetoric of this emerging world of membrane research was seen as another form of proof. As one interviewee said to me, "Given that someone from the university is now describing its history, membrane technology must be something to take into account" (Van Lente & Rip, 1998).

Ultimately, I could articulate that technology does not start with a problem but rather a promise. In contrast to the standard notion of engineers identifying problems to be solved, I now saw engineers inventing an imaginary world. And they do what they do in order to realize this imaginary world. The reader may ask: "Have I heard right? Is human life in its most human dimension a work of fiction"? Yes, now I was able to phrase how the work of engineers begins with fiction—how it is embedded in fiction, assessed by fiction, and propelled by fiction.

## The Past of Futures

Ortega's *Meditación de la Técnica* does even more. After investigating life as drama, as necessity conditioning all other needs, Ortega dwells on the rise of technology in modern societies. Here, he roughly distinguishes between three modes or phases in which technology appeared in the history of Western societies. The chronology Ortega suggests is not original and too cursory for historical purposes, but he introduces a conceptual twist that, again, helped me in my thinking about how expectations are part of technological change. In the first phase Ortega identifies, technology is fully situated in everyday life: Homo sapiens use tools, and the skills for doing so are by now common practice. When humans began to urbanize, special skills were needed and technologies became part of craftsmanship. Some specialized as blacksmiths, others as carpenters or architects. The appearance of modern technology in the third phase was marked by the founding of engineering schools in the late eighteenth century. Here, Ortega argues, the continuous improvement of technology was postulated as a possibility. Of course, the idea of technological progress was not entirely new, but it had been seen as a token of good craftsmanship. When craftspeople successfully used and improved the technology of their trade, it indicated the high quality of their work, establishing their reputation as blacksmiths, carpenters, or architects. Technology was hidden behind the person, as Ortega phrases it. In the third phase, however, technology itself appeared: The notion of "technological progress" became visible and it was the task of engineers to take care of it.

What I find appealing in this reflection is the idea of guaranteed progress: the notion that technological improvement can rightfully be expected to occur. The engineer is certain to find a novel solution, Ortega notes, but how can that be? What strikes me about this exercise is Ortega's intellectual audacity: Instead of presenting a historical reconstruction, he highlights an idea, a principle. The reduction of the history of technology to three phases may be too facile, but it helps to articulate the character of modern technology. It is no longer hidden behind craftsmanship—instead, technology refers to human beings' confident assumption that technical solutions will be found—continuously, now, and in the future.

Ortega's idea of modern technology as the certainty that improvements will occur, points to an overarching generic promise. In my fieldwork, I encountered such generic promises when spokespersons of a particular technology—say membrane technology—embedded their claims in the idea of technological progress as such. Their reasoning, then, was that when it is certain that technological progress will occur, this particular technology is a likely candidate to facilitate it. Their claims found fertile soil since decision makers in firms and policy circles also departed from the conviction that *some* technology must be promising. Their sole task was to decide which one it was. The importance of the generic promise inspired me to elaborate on how technological promises are *nested*: Smaller promises (of, say, a new material) refer to more encompassing promises of a particular technology, which, in turn, are eventually supported by the generic promise—that is, the culturally embedded conviction that there will be technological progress.

The modern certainty that there will be progress is not just a matter of ideology; it is paralleled by a societal task division. I coined this as the *mandate* for engineers: When society in general adopts the idea that technology brings progress, engineers are "mandated" to decide which directions and options are promising. That is, engineers are appointed as the rulers of technological promise, and this comes with both privileges and obligations. On the one hand, this mandate implies a freedom to decide on behalf of others what the next technological promise will be and what is worthwhile to pursue. Is biotechnology the next big thing or is it artificial intelligence? Engineers and technological prophets will inform us. On the other hand, it also brings accountability: Engineers are expected to take good care of technological promise—again, on behalf of others. And when promises appear to fail or when other concerns about technology emerge, the efforts of engineers will be judged. When, for instance, a country is not as rapid as other countries in pursuing a promising technology, the question will be: Who is to blame? Were the engineers sufficiently motivated to keep up with progress? Have they neglected the promise? What interested me was not the particular outcome of the blame question but the very idea of blaming itself. It refers to a moral shortcoming and to a frustrated, unfulfilled expectation that progress should be inevitable.

## Conclusion: The Merits of a Detour

My research on the role of expectations in the development of technology developed into a so-called sociology of expectations, which describes what it means that engineers do not begin with problems to be solved but promises to be fulfilled. The basic tenets are that innovations take place in a "sea of expectations" (Van Lente, 2012). Firms and engineers position themselves in an imaginary future world and act accordingly. They try to decide what the promising routes are and use the actions of their competitors as indications. Is membrane technology the future? If so, we cannot wait, and we should join the bandwagon. Others, in turn, seen their actions as proof that this is indeed the way to go—clearly, this is a self-fulfilling prophesy.

A sociology of expectations provides a novel approach to the question of direction in the study of innovation (Konrad et al., 2017). In this chapter, I reconstructed how this novel approach has benefited from an unexpected approach. I showed how a decisive turn in my research did not come from the usual suspects in my field but from an early twentieth-century philosopher who based his thinking on the vitality of the human spirit (Dust, 1991). Yet his emphasis on the constitutive role of the "good life"—the invented life that fuels human actions—provided the research on expectations with a novel twist. It reversed the gaze on expectations.

My account also testifies to the merit of intellectual excursions in general. The unfamiliar phrasing, the unexpected queries, and a surprising style of reasoning can offer a novel approach to research puzzles. In my case, it brought some new insights, a unique perspective, and a bold way of reasoning. Interdisciplinary excursions may shift the angle, provide a new vocabulary, and, in this way, sharpen concepts. My account also shows that such gains cannot be planned. While not all excursions may be useful, it certainly pays off to step outside of one's research plan every now and then and enjoy the detour.

# References

Dosi, G., Freeman, C., Nelson, R. R., Silverberg, G., & Soete, L. (Eds.). (1988). *Technical change and economic theory*. Pinter.

Dust, P. (1991). Freedom, power, and culture in Ortega y Gasset's philosophy of technology. *Research in Philosophy and Technology, 11*, 119–153.

Fagerberg, J., & Verspagen, B. (2009). Innovation studies—The emerging structure of a new scientific field. *Research Policy, 38*(2), 218–233. https://doi.org/10.1016/j.respol.2008.12.006

Konrad, K., Van Lente, H., Groves, C., & Selin, C. (2017). Performing and governing the future in science and technology. In U. Felt, R. Fouché, C. A. Miller, & L. Smith-Doerr (Eds.), *The handbook of science and technology studies* (4th ed., pp. 465–493). MIT Press.

Kuhn, T. S. (1962). *The structure of scientific revolutions*. University of Chicago Press.

Nelson, R. R., & Winter, S. G. (1977). In search of useful theory of innovation. *Research Policy, 6*(1), 36–76. https://doi.org/10.1016/0048-7333(77)90029-4

Ortega y Gasset, J. (1961). Man the technician. In J. Ortega y Gasset (Ed.), *History as a system and other essays toward a philosophy of history* (pp. 85–161). Norton. (Original work published 1930).

Pinch, T. J., & Bijker, W. E. (1984). The social construction of facts and artefacts: Or how the sociology of science and the sociology of technology might benefit each other. *Social Studies of Science, 14*(3), 399–441. https://doi.org/10.1177/030631284014003004

Rock, P. E. (1979). *The making of symbolic interactionism*. Palgrave Macmillan.

Rogers, E. M. (2003). *Diffusion of innovations* (5th ed.). Free Press.

Simon, H. (1957). *Models of man, social and rational: Mathematical essays on rational human behavior in a social setting*. Wiley.

Van Lente, H. (2000). Forceful futures: From promise to requirement. In N. Brown, B. Rappert, & A. Webster (Eds.), *Contested futures. A sociology of prospective techno-science* (pp. 43–64). Ashgate Publishing Company.

Van Lente, H. (2012). Navigating foresight in a sea of expectations: Lessons from the sociology of expectations. *Technology Analysis & Strategic Management, 24*(8), 769–782. https://doi.org/10.1080/09537325.2012.715478

Van Lente, H., & Rip, A. (1998). The rise of membrane technology. From rhetorics to social reality. *Social Studies of Science, 28*(2), 221–254. https://doi.org/10.1177/030631298028002002

# A Modernization Perspective on Dutch Universities in the Nineteenth Century: Theoretical Sociology Challenging Historiography

Joseph Wachelder

## Introduction

When engaging in interdisciplinary research, conceptual encounters may occur for different reasons and emerge in different phases of a given project. In the case discussed here, a project on which I worked in the late 1980s, unforeseen conceptual encounters and even conflicts presented themselves in the course of my investigations and, in particular, in the project's results. When I now reflect on research I did as a young, inexperienced scholar, I realize that a number of conditions were in place that can explain the unexpected results. These conditions are often connected to interdisciplinary work. This kind of work tends to start from a real-life problem, involving stakeholders with different perspectives on the topic of investigation. Furthermore, one will often find a collaborative spirit

I would like to thank Ton Brouwers for editing this chapter.

J. Wachelder (✉)
Department of History, Maastricht University, Maastricht, The Netherlands
e-mail: jo.wachelder@maastrichtuniversity.nl

between scholars from various backgrounds, while their interests and focus may overlap. Another relevant factor in this kind of work is ample room for experiment in combination with the availability of exemplary studies that can guide the way. These features defined the circumstances that gave rise to conceptual conflicts at the end of my project. In this chapter, I will reconstruct how these conceptual encounters came about, as well as how they became productive and how eventually they were resolved.

Many readers of this chapter will have an academic background, and they may even work in academia or perform research in other contexts. Most academics are familiar with the lofty ideals that in the past were attributed to academic life, as regards both research and education. As of the early nineteenth century, the concept of *Bildung*, as well as Wilhelm von Humboldt's notions of *Einheit von Forschung und Lehre* and *Lehr- und Lernfreiheit*, became increasingly part of how universities defined their teaching and research tasks. If many academics today will respect the era to which these notions belong, they will also consider it a bygone period—a past that is definitely over. At the same time, there has been a steady increase in complaints about the directions of and developments in today's academic life. Managers and administrators of universities tend to concentrate their efforts on steering their institutions into new directions by formulating a never-ending train of mission statements, strategies, and policies. More often than not, however, their effectiveness and desirability instantly meet with skepticism from faculty staff, often referring to the need for *Bildung* again.

As a young scholar I studied the relevance of academic ideals, missions, and strategies in Dutch higher education in the nineteenth century. Below, I will first introduce the real-life issue that made me embark on this interdisciplinary research project, without realizing its intricacies. Next, I will introduce the theoretical framework that guided my research, and I reflect on the different motivations underlying the decisions involved, one of which pertained to the interdisciplinary team in which I was embedded. For my analysis of educational debates and educational philosophies in nineteenth-century Dutch higher education, I used the sociological theory of Niklas Luhmann as a frame. Another important study I relied on was Rudolf Stichweh's 1984 book on the emergence of the system of disciplines in Germany, entitled *Zur Entstehung des modernen Systems wissenschaftlicher Disziplinen: Physik in Deutschland 1740–1890*. Subsequently, I will consider my findings, as well as observations that—in retrospect—were surprisingly absent in my study, which pertains in particular to Humboldt's idea of a university in nineteenth-century debates about Dutch higher education. Further reflections on these discrepancies will feed into the conclusion, where I demonstrate that starting from a research question that engages stakeholders and is based in actual practices may give rise to unexpected results precisely because one does not start from canonized disciplinary knowledge. My closing argument will be that an interdisciplinary approach is not necessarily more encompassing or comprehensive than a disciplinary one.

## A Contested Mission Statement Feeding into a Research Project

Maastricht University (UM) is a fairly young, state-funded university, established in 1976 (Klijn, 2001, 2016; Knegtmans, 1992). A major reason for the Dutch government to grant the Province of Limburg a new *Rijksuniversiteit* was tied to the region's dire economic situation after the closure of the region's state-operated coal mines. The intention to implement a fresh approach to university education further motivated the government's decision to establish a new academic facility. From its inception, State University Limburg, as used to be the name of the UM until 1996, embraced the principles of Problem-Based Learning (PBL), as developed by McMaster University (Hamilton, Ontario). Because the UM started out as a medical school, this educational philosophy was a perfect fit, as McMaster first introduced this PBL approach at its medical school in 1969 as well. Next to this educational philosophy, the UM's mission statement entailed an alternative approach to medical care, prioritizing primary healthcare rather than state-of-the-art medical interventions in an academic hospital. This last aspect actually became an issue when the Medical Faculty grew rapidly and its high-tech departments started to flourish and sprawl. The didactic principles of PBL grew more contested after the founding of new faculties, such as the Faculties of Law and Economics. The UM's Executive Board wanted the new faculties to embrace the educational principles of PBL as well. But critics voiced concerns about its usefulness outside the pragmatic, hands-on domain of medicine. And even staff within the Medical Faculty had reservations about PBL because of its explicitly student-centered approach, which required staff first and foremost to assume a service-oriented role that might go at the expense of goals linked to research and theoretical knowledge development.

To guide these discussions—and/or to appease the most radical promoters of both positions—the UM's executive board, supported by the faculty deans, decided to open up a position for someone to organize an ongoing intellectual debate on UM's educational philosophy. In combination with historical research on the role of educational philosophies and innovations in nineteenth-century Dutch higher education, this project was meant to create some common ground. Discussions would keep the potential relevance of educational philosophies alive, regardless of the outcomes. I seized the opportunity to devote my research project to educational debates in nineteenth-century Dutch universities. At the same time, I was quite aware of the highly normative positions and fierce debates frequently triggered by educational missions and philosophies.

The engagement of stakeholders affected my research project. The educational debates going on at Maastricht University in the 1980s determined both the leading question and the unit of analysis of my historical research. Interdisciplinary research is often promoted by arguing that real-life problems transcend disciplinary boundaries. Yet, real-life problems are not a given; stakeholders need to articulate them. For the type of interdisciplinary research I performed, the engagement of stakeholders in delineating the central research interest was key. Meanwhile, new concepts such as transdisciplinary research and community-engaged research have been introduced, to highlight the benefits of engaging stakeholders in research, not just in articulating the research question but throughout the research trajectory.

## An Interdisciplinary Research Environment

My research project grew almost naturally into an interdisciplinary affair. In the project, I addressed the educational discussions going on in Maastricht University's Faculties of Medicine, Public Health, Law, and Economics. At the time, the Humanities faculty was under development, meaning that it was not yet formally established. In this context, a group of young scholars from different disciplinary backgrounds shared offices in a provisional setting, including theoretical sociologists, sociologists of law, philosophers, philosophers of science, historians, and science and technology studies (STS) scholars. As there was no narrowly defined institutional frame yet, their interactions were intellectually open and diverse. Moreover, the supervisors of my research project, Louis Boon and Jeroen Dekker, gave me a lot of freedom to explore relevant issues, while also protecting me from getting lost along the way. Louis Boon was trained as a psychologist to become a philosopher with an interest in evolutionary models of science (Boon, 1983). Jeroen Dekker, a historian of pedagogics and education, had an interest in long-term cultural changes (Amsing et al., 2018) and taught me, trained as a theoretical physicist, the finesses of historical source analysis.

The design, theoretical framework, and methodology of my research were highly affected by the social environment sketched above (Wachelder, 1992). Given my focus on nineteenth-century educational debates, I relied on publicly available sources, such as inaugural lectures and political debates. I decided to compare and contrast debates about the character and legislation of higher education in general, with educational debates in the faculties of Medicine and Law mirroring the contemporary situation at Maastricht University. My theoretical framing was inspired by the interactions and disputes within a close circle of colleagues, and these efforts concentrated on bridging the actor/structure divide in sociology and the role of contingency in social and natural processes. Tannelie Blom, the late Werner Callebaut, Ton Nijhuis, Nico Roos, and I met frequently in a reading group to discuss a variety of texts and books from different disciplines. Among other things, we made a meticulous study of Niklas Luhmann's *Soziale Systeme* (1984), which would affect the intellectual career of all of us (Blom, 1997; Callebaut, 1993; Nijhuis, 1996; Wachelder, 1992), even though we would, later, hive off into different directions.

Three aspects of Luhmann's approach of self-organizing social systems seemed apt in particular for analyzing educational debates on modernizing Dutch universities in the nineteenth century. First, by considering communications as the basic elements of social systems and social structures as double contingent expectations of expectations expressed in discourse, Luhmann got rid of deterministic tendencies in systems theory. This offered a consistent and coherent perspective to bridge the actor/structure dilemma in theoretical sociology. Secondly, Luhmann's theory of self-organizing social systems conceived modernization as an overall transition from static (hierarchical) organizational principles to focused (functional) process-oriented ones, which allows for an increase in social complexity. Third, these subsystems are considered to function rather independently from each other; the interactions among them are conceived as "interpenetration," as "noise," causing them to interact without determining each other or assuming a hierarchical relationship between different subsystems (Luhmann, 1984).

Of course, one may have doubts about the benefits of this rather abstract social theory for the study of educational debates in Dutch universities, as is also testified by the argument in van Leeuwen's chapter in this volume. However, not prioritizing one subsystem over others allows for studying their mutual interactions as concrete manifestations in space and time. The theoretical conceptualization of communications as the basic elements of social systems and discourses as structures offers a sound foundation for using discourse analysis to study changes in society. Luhmann authored many books and articles using discourse analysis to study the transition from traditional to modern societies for specific domains. His interpretation of modernization as a social transition from rather static, hierarchical organizational principles to process-oriented ones would serve me as a useful lens to study the nineteenth century.

Rudolf Stichweh's study (1984) on the emergence of the system of disciplines, which also started from Luhmann's social systems theory, provided a decisive push to use it as a framework for studying debates about and within nineteenth-century Dutch education. If interdisciplinary research will often be seen as an alternative for studies framed along disciplinary lines, historically informed readers will immediately acknowledge that the disciplinary organization of science only came about in the nineteenth century. Stichweh explains its emergence as involving a long-term transition from a hierarchical organization of knowledge or studies to a processual one. In this context, the notion of hierarchy refers to faculties or given methods. The disciplinary organization allowed more freedom in determining research interests and selecting appropriate methods. Its processual character was supported by the emergence of scientific journals, which fueled and speeded up scholarly exchanges on focused and thus more limited topics. At the time, I was impressed by Stichweh's study (and I still regard it as one of the best accounts on discipline formation), and I said to myself: why not give Luhmann's social systems theory a try as a frame for analyzing educational debates and interpreting changes in nineteenth-century Dutch higher education?

## Identifying and Interpreting Debates in Nineteenth-Century Dutch Higher Education

A scholarly dive into nineteenth-century sources on higher education is likely to produce a compassionate smile on many a scholar's face. At the time, as well as today, scholars and professors did not only debate endlessly—which is also a major task of their job, of course—but they also complained vehemently about the status quo of higher education. At first sight, many of the topics they covered are quite recognizable for us today, if not very much the same. Professors lamented each and every threat that might raise their workload, and they had concerns about courses that they were either expected to teach or not allowed to teach. Likewise, complaints about the lack of passion or motivation among students appear to be a recurring element in academia. Discussions about desirable measures, legislation, or the financing of universities by government show many similarities and continuities over 200 years as well. Although the king or government set up many special committees that would engage in lengthy debates on all sorts of issues, only a few proposals were turned into legislation. Many implemented measures were rapidly discontinued again because of the ill-considered or detrimental effects they produced. For instance, the introduction of an entrance exam for universities led to a sharp decline in student enrollments. Should we conclude, then, that there is always something to complain about and that professors are just ordinary human beings?

Despite recurring manifestations of academic displeasure and frustration, much has changed in their content, context, and connections, even when disregarding its media and ways of display. For one thing, nineteenth-century academics were largely grumpy old men. The first female student at a Dutch university, Aletta Jacobs, entered the Medical Faculty of the University of Groningen not until 1872, while the first female professor, Johanna Westerdijk, was inaugurated only in 1917 (Bosch, 1994, 2005). But among the exclusively masculine professoriate, there were also many who saw a need to leave the beaten track. The modernization perspective derived from Luhmann's social systems theory was helpful to interpret what was at stake in different educational debates, compare and contrast educational debates in different faculties, and study links between as well as to explain connections between the manifold concerns and worries on minor or major educational issues, ranging from practicalities to educational philosophies.

To demonstrate the entanglement of a seemingly minor issue with other debates, the dispute about the freedom of students to determine their own order of study at universities offers a good example (Wachelder, 1991). Academics who were concerned about the performance of a large segment of the university student population came up with simple solutions, such as prescribing the order of studies and more serious exams. Both proposals regularly recurred in the nineteenth century, and they would frequently meet with fierce resistance. Students' freedom to arrange the order of their study at a university predated the nineteenth century; it was not only considered a privilege, but it also served a specific educational aim. It would help them later in life, when holding a responsible position in society, which required an independent attitude and autonomous judgment. Moreover, from an institutional point of view, freedom of study distinguished Latin schools (called "gymnasia" later on) from universities.

To appreciate the distinguishing characteristic of freedom of learning, the fact that higher education comprised both Latin schools and universities is crucial. Latin served to connect the two educational institutions, and proficiency in Latin was considered a prerequisite for those in the upper classes. Compared to other countries, universities in the Netherlands held on to Latin as lingua franca rather long. Latin schools proceeded with annual, successive classes. Education at universities should be free. What else could serve as distinguishing characteristic in higher education? In particular, professors who taught the first propaedeutic part, such as Philip Willem van Heusde (1778–1839), defended the freedom of study at universities fervently. Moreover, not all academic courses required a final examination; for some a testimonial sufficed. That professors did not receive a set monthly salary, as their payment depended on the number of registered students for their course, made the matter all the more intricate.

The ongoing use of Latin in academia impacted the political debate as to how to improve the educational offerings in the Netherlands. The development of industry and commerce required more and new competences from citizens. But how to achieve this? Until 1865, there were only two kinds of facilities of higher education in the Netherlands: gymnasia and universities. Moreover, instead of referring to primary and secondary education, the Dutch used the adjectives "lower" and "middle." In the nineteenth century, a huge debate emerged as to what this "mid-level" education, at that point still to be established, should entail. Some argued that gymnasia should devote more attention to the natural sciences. Yet, Latin schools held on to their mission: training the (administrative and scholarly) upper classes, for which classical languages were deemed essential. In 1865, new school types for secondary ("middle") education were introduced, geared to jobs in commerce and industry rather than preparing young men for university. Yet, the level of teaching in the natural sciences at the newly established *Hogere Burger School* (HBS, or "civic high school") set a standard that the gymnasia had trouble meeting. Within less than two decades, the gymnasia had to comply with the new standard set by the HBS in teaching the natural sciences.

In hindsight, one can interpret the debates and discussions in conjunction with the overall transition in education from an organization based on static principles, involving a privileged class and a fixed distribution of professional roles, to a meritocracy, with a focus on exams and the individual's learning process. This kind of sociological perspective was lacking in the scholarly reflection on higher education at the time because its very establishment was part of the same transition. Moreover, the intricate relationships between many, seemingly disparate issues made it difficult to develop a comprehensive overview. Many contemporary diverging arguments made sense, at least on paper. At the same time, new educational institutions, such as the HBS system, would develop into directions different from the ones originally envisaged. The educational debates going on in medical and law faculties at the time partly tapped into the developments described above, showing an increasing orientation toward processes as well.

In contrast to the mostly implicit references to processes as alternative organizational principles in general debates about higher education, explicit references to processes were made in Dutch medical faculties. In the beginning of the nineteenth century, a wide variety of medical practitioners was active, aside from academically trained physicians. Controlled by government at the provincial level, there were different training trajectories for medical practice, leading to a variety of licenses for specific sectors of medical practice, whereby it was common to distinguish between medical practice in cities and in the countryside (Van Lieburg, 1983). In 1826, clinical schools were established, to replace practice training by masters, which used to complement medical education at universities. Pleas for a nationwide surveillance of medical practitioners were accompanied by arguments to abandon the many peculiar distinctions in what ought to be a unified profession. Diseases ought to be considered as processes that aren't limited to specific parts of the body or types of intervention.

The emphasis on diseases as processes went along with underlining physiology, the study of processes of life, as the overarching basis of medical knowledge and practice, rather than nosological classifications and systems based on static hierarchies. The envisaged transition produced significant discursive misunderstandings, for instance, about the meaning of "experience." Some started to distinguish new experience, as based on experiment, from old experience, as based on practice. The process- and science-based approach to medicine necessitated, according to its promoters, new educational formats, such as microscopy classes and chemical and physiological educational labs, to train the observational skills of medical students. Needless to say, no evidence-based educational research was available to support these claims. Essentially, the educational formats proposed were derived from changing preferences as regards their content.

In the 1860s, pleas in the medical faculties promoting more and different education in the natural sciences to improve physicians' observational skills began to interfere with proposals for educational change on a more general level. Experts saw more and stricter exams, organized on a national scale, as a crucial element. In 1865, state exams in both the natural sciences and medical practice were introduced. This led to the almost immediate discontinuation of clinical schools, which could not keep up with the high level of training required for the natural sciences. The introduction of a new Law on Higher Education, in 1876, triggered a debate on whether graduates from the newly established HBS could have access to the study of medicine, even though this was not part of the original reason for setting up the HBS system. After fierce debates in the House of Representatives, HBS graduates were considered admissible if they passed an additional entrance exam for Latin and Greek. Only two years later this requirement was abandoned again. The increased relevance of the natural sciences for medicine turned out to be the decisive factor.

Where change-minded medical professors highlighted the process-character of disease, to motivate national government officials to intervene in the training and surveillance of medical practitioners, as of 1838 law faculties had to address the new Civil Code, in the wake of several tumultuous decades in which the French Civil Code was upheld. Some feared that the new codes of law would make legal study obsolete due to the increased accessibility of the new law books. Although legal study hardly became a superfluous field of study, the new Civil Code stirred debates about the curriculum. Of old, many students of law opted for administrative positions. Law was a type of general-career study also at that time, be it mostly for the elites only. Law professors had turned their field into a highly theoretical endeavor, however. Before the Codification, Roman Law served as subsidiary law, it being taught as the *nec plus ultra*.

Roman Law provided a systematic organization of legal sources, including definitions that could be logically dissected and applied. In the eighteenth century, natural law, with its focus on legal principles, challenged this central position of Roman law in the curriculum. Both competing legal systems, however, shared a focus on formal logic, consistency, and coherency.

With the new Civil Code, the orientation on fixed legal sources or principles seemed to lose its relevance. The position of Roman Law in the curriculum came under debate. Wouldn't an introduction in current law be a more suitable beginning of the curriculum? In the ensuing discussion, the arguments highlighting educational merits of Roman Law came with some remarkable twists, showing the versatility of didactic arguments. A closed, complete, consistent, and coherent law book was unattainable because all sorts of arrangements, such as fiscal ones, were prone to changes. For some, this was an argument to maintain a central role for Roman Law in the curriculum, given the completeness of its sources and its logical, systematical interpretation. Others argued, however, that the development of law, and its changing social context, deserved more attention. New topics such as politics and statistics were introduced into the curriculum. Yet, the focus on the dynamics of law in its concrete social context did not make Roman Law superfluous. New, historical approaches and interpretations of Roman Law emerged that made it an eminent subject to understand the development of law in general. As in medical faculties, the new focus on processes prompted debates and confusion about what practical education could and should entail. Equating theory with systematics and logical analysis did not work any longer; nor did equating practice with experience.

Nineteenth-century educational missions and principles referred, implicitly or explicitly, to a wide variety of social, educational, and scientific aspects that were intricately connected. Few of the positions were backed up with compelling arguments or irrefutable evidence. Some were based on traditions and established knowledge. Others were more forward-looking, pointing to an uncertain future. Some of the traditional arguments given for specific educational methods or content could be easily tinkered with to suit new, changed circumstances.

To identify nodes in the myriad of ongoing debates and to interpret their connections, Luhmann's perspective on modernization proved highly useful in guiding my historical research. In the course of my project, I contacted many senior colleagues, in specialized fields: historians of education; historians of universities; and historians and sociologists of science, medicine, or law; philosophers of education; philosophers of science; even scholars working in public or business administration. I was surprised by the willingness of many of them to share insights or to elucidate subtleties of their research. Conversely, collaboration requires an open, inquisitive mind on the part of the researcher and gratefulness to those who share knowledge and insights. Only occasionally, I encountered mechanisms of exclusion in relation to my decision to ground my research in a sociological theory or publicly accessible sources (rather than archival sources), as exemplified by the rejection of a paper for a historical conference. If disciplinary boundaries can be transcended, they can also be defended as a way to maintain settled assumptions.

## In Search of Wilhelm von Humboldt

Luhmann's perspective on modernization helped me to identify and interpret nineteenth-century educational debates in the Netherlands. Yet, I did not come across references to the neo-humanist scholar Wilhelm von Humboldt (1767–1835) in nineteenth-century debates about Dutch higher education. Humboldt has gained worldwide acclaim for his ideas on *Bildung*, as associated with the establishment of Berlin University in 1810 and the overall German university model. His ideas of *Einheit von Forschung und Lehre* and *Lehr- und Lernfreiheit* have become canonized elements of how universities conceive of their identity, also in the Netherlands. The legislation of Dutch Higher Education in the Netherlands is marked by two major laws, the so-called Organiek Besluit from 1818, redressing the French occupation, and the Law on Higher Education from 1876. The last one is traditionally connected with the name of Humboldt because the law explicitly singles out research as one of the main tasks of universities (Wachelder, 2001). The introduction of research in nineteenth-century universities was, and partly still is, associated with the German university model, which aside from education defined research as a main task of universities, while also highlighting the interconnection of teaching and research. The German model of a university is said to have been imported to many other countries, including the Netherlands. Although I ran into the notion of "freedom of study" in my investigations at the time, I did not see it linked to Humboldt. Moreover, the meaning and connotations of *Lernfreiheit* in the early nineteenth century differed from the lofty academic ideals associated with Humboldt.

That I missed the link with Humboldt's views forced me to reflect on my results. Did Luhmann's lens of modernization perhaps make me focus so much on a transition from static hierarchies to processes that it blinded other relevant discussions? Did I overlook relevant sources? I started to systematically study the acts of the House of Representatives regarding the Law on Higher Education from 1876 and rather unknown sources, such as Robert Vorstman's book on German universities and their histories (Vorstman, 1872). I found hardly any references to Humboldt (Wachelder, 2003). Increasingly, it became clear how my research (Wachelder, 1992) tapped into an ongoing international reassessment of Humboldt's importance for the renewal of German universities in the nineteenth century. This reevaluation of Humboldt's impact on nineteenth-century German universities or a German university model extended across many decades in which scholars successively addressed different aspects (Wachelder, 2003).

The dominant interpretation of Humboldt's educational philosophy, championing individual scholarship, became challenged as of the late 1970s (Lechner, 2003). In the 1980s, historical studies placing Humboldt and German universities in their social context set the tone (Labrie, 1986; McClelland, 1980). Moraw (1984) concluded from his meticulous case study of the University of Gießen that Humboldt's ideas of scholarly isolation—*Einsamkeit und Freiheit*—had less impact than his short stay at the Ministry of Inner Affairs. In that position, Humboldt advocated that the government should have the final say in the appointment of new professors. It effectively abolished the "family university," where sons succeeded their father, without a serious assessment of their competences. As regards the idea of *Lernfreiheit* (freedom of study), Moraw (1984) came up with the telling observation that advocates of innovative teaching laboratories in the natural sciences often opposed the idea of freedom of study and favored an organized curriculum.

Schubring (1991) put the reception history of the "Humboldtian model" on the scholarly agenda. He observed that many of Humboldt's ideas were not considered new at all by his contemporaries. Moreover, only in 1900, Humboldt's key publication on the structure and organization of the university in Berlin—*Über die innere und äußere Organisation der höheren wissenschaftlichen Anstalten in Berlin*—was published (Paletschek, 2001, p. 76). Adolf von Harnack's history of Berlin University dates from 1910 and was published on the occasion of the university's centenary. That year also saw the publication of Berlin University's *Gründungsschrifte*, comprising contributions by Wilhelm von Humboldt, Friedrich Schleiermacher, Johann Gottlieb Fichte, and Henrik Steffens. As concluded by Sylvia Paletschek (2001, p. 77), the notion of a Humboldtian university model dates back to the 1920s, but this notion developed into a topos not until after the World War II, in particular, in the 1960s. In 1997, the German translation of Mitchell G. Ash's edited volume reevaluating German universities in the past and future was entitled *Mythos Humboldt: Vergangenheit und Zukunft der deutschen Universitäten* (Ash, 1997).

In hindsight, then, it is obvious that the concept of a Humboldtian or German model was hardly useful for interpreting the development of Dutch universities in the nineteenth century (Wachelder, 2001). In 1992, my application of Luhmann's modernization perspective on nineteenth-century Dutch educational debates challenged established historiography and the concept of a German university (Wachelder, 2001). Starting from a research interest in educational debates, comparing and contrasting those on different levels and in different faculties, while applying a theoretical framework based on theoretical sociology, turned out helpful to put aside twentieth-century topoi in university's self-descriptions. Unintendedly, the results of my study questioned the concept of a German university model, fueling research-oriented universities.

# Conclusion: Foregrounding and Backgrounding

As my case study reveals, conceptual encounters presented themselves mainly during the stage of reflection on the study's findings rather than in the research process. These conceptual encounters, I argued, were produced in particular by starting from an authentic research question, informed by real-life issues, largely raised by stakeholders. Real-life challenges did not only inform my research question but also influenced its major unit of analysis and the sources studied. The above argument aligns with arguments often heard in promoting interdisciplinary research. Such research is needed, some argue, to analyze real-life problems and suggest practical solutions, given that both transcend disciplinary boundaries. Another argument frequently brought up for promoting interdisciplinary research is that multi- or interdisciplinary work leads to more comprehensive results than disciplinary research. I doubt whether this applies to the research project described above, however. For one thing, it is questionable whether completeness is an asset per se in an information-saturated world. My interdisciplinary work benefited in particular from an innovative research question, as well as an unconventional research design and method.

Over the last two decades, I focused my research no longer predominantly on education or the history of universities. Yet, I kept abreast of the field, among other things, by reading and reviewing many new publications. From which ones did I learn most? Rather than highlighting comprehensive histories of universities, I would like to highlight two books, which have in common that they adhere to a well-chosen and underexplored unit of analysis.

Remieg Aerts's biography of Rudolph Thorbecke complements my project in two principal aspects (Aerts, 2018). Whereas I used the lens of modernization to focus on transitions from hierarchies to processes, Aerts zeroes in on the establishment and further development of a constitutional Kingdom in the Netherlands as of 1848. First and foremost, this directs his analysis to the spatial dimension, in particular the level of the national government, at the expense of the regulating power at the provincial level. Second, his approach reveals the political logics behind some awkward phenomena I came across. In my analysis, the 1863 Law on Secondary Education and the 1865 Law regulating the admission to the medical profession anticipated and highly determined the outcome of new legislation on higher education in 1876. Historians of universities tended to disregard this prior development and zoomed in on the 1876 Law on Higher Education. Aerts discusses, in great detail, Thorbecke's contributions to three cabinets: 1849–1853, 1862–1866, and 1871–1872. From a political perspective, it made sense to propose new legislation for primary, secondary, and higher education in that particular, consecutive order.

The second publication that I would like to put in the spotlight complements my study in a different way. By concentrating on Dutch student periodicals from the nineteenth century, Annelies Noordhof-Hoorn (2016) gives students a voice, an element that is lacking in my 1992 study. Giving voice has a literal meaning in this context: in the eighteenth century, student periodicals still hardly existed. Metaphorically, giving voice here implies that in the nineteenth century, students at Dutch universities became increasingly critical, initially with regard to the specific education provided to them and, later on, concerning education in a wider social context. Noordhof-Hoorn's analysis of the production of these student periodicals reveals the impact of infrastructural works, in particular bridges and railways, on student recruitment and student mobility. Rather than serving local, regional, or provincial interests, Dutch universities developed into national institutions in the course of the nineteenth century.

I address the merits of interdisciplinary work preferably in terms of foregrounding innovative research questions, research designs, and methodologies to which the engagement of stakeholders may contribute significantly. This comes at the price of backgrounding disciplinary assumptions and logics. In my experience, having the privilege of collaborating with inquisitive colleagues from whatever discipline not only broadens the scope but also deepens the analysis of a research project.

# References

Aerts, R. (2018). *Thorbecke wil het: Biografie van een staatsman.* Prometheus.

Amsing, H., Bakker, N., Van Essen, M., & Parlevliet, S. (Eds.). (2018). *Images of education: Cultuuroverdracht in historisch perspectief.* Passage.

Ash, M. G. (Eds.). (1997). *Mythos Humboldt: Vergangenheit und Zukunft der deutschen Universitäten.* Böhlau.

Blom, T. (1997). *Complexiteit en contingentie: Een kritische inleiding tot de sociologie van Niklas Luhmann.* Kok Agora.

Boon, L. (1983). De list der wetenschap. Variatie en selectie: vooruitgang zonder rationaliteit. Ambo.

Bosch, M. (1994). *Het geslacht van de wetenschap: Vrouwen en hoger onderwijs in Nederland 1878–1948.* SUA.

Bosch, M. (2005). *Een onwrikbaar geloof in rechtvaardigheid: Aletta Jacobs 1854–1929.* Balans.

Callebaut, W. (1993). *Taking the naturalistic turn: How real philosophy of science is done.* The University of Chicago Press.

Klijn, A. (2001). *Onze man uit Maastricht. Sjeng Tans 1912–1993: Een biografie.* SUN.

Klijn, A. (2016). *Het Maastrichts experiment: Over de uitdagingen van een jonge universiteit 1976–2016.* Vantilt.

Knegtmans, P. J. (1992). *De Medische Faculteit Maastricht: Een nieuwe universiteit in een herstructureringsgebied, 1969–1984.* Van Gorcum.

Labrie, A. (1986). *'Bildung' en politiek, 1770–1830: De 'Bildungsphilosophie' van Wilhelm von Humboldt bezien in haar politieke en sociale context.* Historisch Seminarium van de Universiteit van Amsterdam.

Lechner, D. (2003). In eenzaamheid en vrijheid? Wilhelm van Humboldts idealen voor de universiteit. *Jaarboek voor de geschiedenis van opvoeding en onderwijs, 4,* 71–89.

Luhmann, N. (1984). *Soziale Systeme: Grundriß einer allgemeinen Theorie.* Suhrkamp.

McClelland, C. E. (1980). *State, society, and university in Germany 1700–1914.* Cambridge University Press.

Moraw, P. (1984). Humboldt in Gießen: Zur Professorenberufung an einer deutschen Universität des 19. Jahrhunderts. *Geschichte und Gesellschaft, 10*(1), 47–71.

Nijhuis, T. (1996). *Over de grenzen van het sociaal-wetenschappelijk verklaringsideaal in de Duitse geschiedschrijving.* Van Gorcum.

Noordhof-Hoorn, A. (2016). *De stem van de student: Nederlandse studentenbladen in de negentiende eeuw.* Verloren.

Paletschek, S. (2001). Verbreitete sich ein 'Humboldt'sches Modell' an den deutschen Universitäten im 19. Jahrhundert? In R. C. Schwinges (Ed.), *Humboldt International: Der Export des deutschen Universitätsmodells im 19. und 20. Jahrhundert* (pp. 75–104). Schwabe.

Schubring, G. (Ed.). (1991). *'Einsamkeit und Freiheit' neu besichtigt: Universitätsreformen und Disziplinenbildung in Preussen als Modell für Wissenschaftspolitik im Europa des 19. Jahrhunderts.* Franz Steiner.

Stichweh, R. (1984). *Zur Entstehung des modernen Systems wissenschaftlicher Disziplinen: Physik in Deutschland 1740–1890.* Suhrkamp.

Van Lieburg, M. (1983). De tweede geneeskundige stand (1818–1865): Een bijdrage tot de geschiedenis van het medisch beroep in Nederland. *Tijdschrift voor Geschiedenis, 96*(3), 433–453.

Vorstman, R. (1872). *De hoogeschool in Duitschland.* Werst (Firma: Jacs. Hazenberg, Corn. Zoon).

Wachelder, J. (1991). Umgestaltung des niederländischen akademischen Unterrichts im neunzehnten Jahrhundert nach einem deutschen 'Modell'? Das Modell: Modernisierung – Das Gymnasium zwischen dem akademischen und dem Sekundarbereich als Beispiel. In G. Schubring (Ed.), *'Einsamkeit und Freiheit' neu besichtigt: Universitätsreformen und Disziplinenbildung in Preussen als Modell für Wissenschaftspolitik im Europa des 19. Jahrhunderts* (pp. 227–254). Franz Steiner.

Wachelder, J. (1992). *Universiteit tussen vorming en opleiding: De modernisering van de Nederlandse universiteiten in de negentiende eeuw.* Verloren.

Wachelder, J. (2001). The German university model and its reception in the Netherlands and Belgium. In R. C. Schwinges (Ed.), *Humboldt International: Der Export des deutschen Universitätsmodells im 19. und 20. Jahrhundert* (pp. 179–204). Schwabe.

Wachelder, J. (2003). Wilhelm von Humboldt en de Nederlandse Wet op het Hoger Onderwijs van 1876. *Jaarboek voor de geschiedenis van opvoeding en onderwijs, 4*, 112–129.

# Constitutional Reform in the Postwar Netherlands: Law in History

Karin van Leeuwen

## Introduction

What is the authority of a constitution when it comes to defining the system of government? In 1948, the Dutch constitutional law professor André Donner compared the state of his discipline to that of 1848, the year of the famous constitutional reform led by law professor and politician Johan Rudolph Thorbecke. Donner observed that times had irrevocably changed since then, because "the unlimited respect for the written constitution is missing, and one hears justified complaints about the respect for constitutional law and its scholarship, complaints also, no less justified, about the unreality of this scholarship itself" (Donner, 1948, pp. 361–2). Notwithstanding these rather pessimistic observations, Donner would, in the next decades, actively contribute to constitutional reforms as a member and chair of consecutive constitutional committees.

K. van Leeuwen (✉)
Department of History, Faculty of Arts & Social Sciences,
Maastricht University, Maastricht, The Netherlands
e-mail: karin.vanleeuwen@maastrichtuniversity.nl

© The Author(s) 2023
K. Bijsterveld, A. Swinnen (eds.), *Interdisciplinarity in the Scholarly Life Cycle*,
https://doi.org/10.1007/978-3-031-11108-2_4

Moreover, as I have concluded from studying the political history of these reforms, Donner's prominent role relied significantly on the constitutional legal arguments he was able to bring to the debate—despite the decreased respect for this discipline he had observed earlier (Van Leeuwen, 2013).

In the initial stages of my research into the history of constitutional reform in the postwar Netherlands, legal writings such as Donner's did not figure prominently. Interested in constitutional politics as a process regarding the content of the constitution, my project primarily focused on the many initiatives attempting to reform the key document "establishing a system of government, defining the power and functions of its institutions, providing substantive limits on its operation, and regulating relations between institutions and the people" (Galligan & Versteeg, 2013, p. 6). As is typical for political history as well as related disciplines interested in the political, law in my project merely featured as the outcome of the process or, more precisely, the result of often difficult negotiations between parties and the ideas, interests, and power they bring into play. By following the paths of various Dutch reform plans—regarding, for example, the electoral system, direct democracy, and demonstration rights—through public opinion, political backrooms, and parliamentary debates, I expected to refine existing explanations for the (lack of) success of these and other reforms, explanations that so far had mostly focused on political calculations and institutional constraints (e.g., Andeweg, 1989).

Soon after, however, I was motivated to delve further into the object of the reforms itself: the constitution. This curiosity was sparked by my systematic analysis of the archives of what we called constitutional committees: temporary committees in which usually politicians as well as legal experts drafted the reforms (Van Faassen et al., 2010). In spite of Donner's observations about the constitution's declining normative value, the debates in these committees suggested that the constitution figured as more than just a blank sheet to be filled as political majorities desired. Rather, as a system of norms and practices, the constitution also appeared to shape the political process by enabling and constraining possible paths for constitutional reform.

Moreover, in this process, the exact meaning of the constitution seemed by no means fixed: various interpretations competed for prominence, in an often-implicit debate underlying the discussions about actual reforms. The Babylonian confusion in which these implicit debates incidentally resulted intrigued me but also left me without the appropriate words to describe them properly—or at least, I had trouble finding them in my own discipline.

In this chapter, I show how my aim to better integrate these legal aspects of constitutional politics inspired me to cross disciplinary boundaries and look for a way to integrate law into political history. In particular, I focus on the concept of (constitutional) tradition(s) that I used to synthesize the various interpretations of the constitution and their normative claims. As I show in the first section, this concept was originally borrowed from constructivist, interpretative political science, but it also tunes into recent innovations in the discipline of legal history within the broader legal domain. The second section then illustrates how discerning three constitutional traditions allowed me to include the legal dimension in my historical narrative of Dutch constitutional reform. Finally, by positioning this example of law-in-history in the expanding interdisciplinary field, the conclusion considers how the concept of tradition might benefit a further integration of law and (political) history.

## Conceptual Explorations: Constitutional "Tradition(s)"

Interdisciplinary research into the politics of constitutional law, or the politics of law in general, has recently generated a significant amount of scholarship (e.g., Versteeg & Galligan, 2013), which this section does not even attempt to summarize. Instead, weaving recent insights through my original explorations, I focus on the concept of tradition and how it aims to cross disciplinary boundaries.

Despite recent moves toward interdisciplinarity, legal scholarship and social scientific approaches to law are still markedly divided by the boundaries described by the French political sociologist Pierre Bourdieu in his seminal work on "the force of law" (1987). As Bourdieu observed, legal studies are typically split into a formalist approach, "which asserts the absolute autonomy of the juridical form in relation to the social world," and an instrumentalist approach, "which conceives of the law as a reflection, or a tool in the service of dominant groups" (1987, p. 814). The former still dominates legal scholarship, serving to construct a coherent body of doctrine. In contrast, the latter approach, which is found both in critical legal studies and in adjacent social science disciplines, seems scarcely interested in the law, which it regards mostly as a cover-up for ideological aims—recent qualifications include "a smokescreen for ideology" (Roux, 2019) or "politics by other means" (Hirschl, 2013).

Bourdieu's legal sociology is only one of many attempts to overcome this hard split. By studying law and its practitioners, to which he referred as the legal field, Bourdieu continued to critically reflect on the social implications of the competition over the right to determine what the law is, as he observed in the more critical approaches. At the same time, he took the relative autonomy of the law seriously, at least as a body of knowledge that provided lawyers with their social capital. A more rigid approach that also focuses on the autonomy of the law—or the self-referential reproduction of legal communication—can be found in Niklas Luhmann's theory of the legal system (*Das Recht der Gesellschaft*, 1993).

Both Bourdieu's strict separation between fields and Luhmann's closed system seemed a rather ill fit for the empirical reality of Dutch constitutional legal debate I aimed to explore. Not only was it nearly impossible to tell where politics ended and law began, constitutional lawyers themselves also questioned the legal character of their discipline, as I will elaborate below. What I did appreciate in these sociological approaches, however, was their genuine interest in the role of legal knowledge and language, as well as cultural capital, in distinguishing lawyers from non-lawyers. A more anthropological interpretation regarded the legal discipline as a distinct culture with a unique manner of making sense of the world (Etxabe, 2020, p. 25; Geertz, 1983). As a political historian formed during a wave of cultural–anthropological approaches to politics (e.g., Te Velde, 1997), I could easily integrate these interpretations in my work.

While sociologists and anthropologists inspired my views on the legal discipline, interpretative work by political scientists offered the key concept that I would use to connect lawyers and their knowledge to their actual contributions to constitutional reform: the concept of tradition. Defined as "a set of connected beliefs and habits that intentionally or unintentionally passed from generation to generation at some time in the past" (Bevir & Rhodes, 2003, p. 34), this concept offered a contingent approach to what other political scientists described as paths or institutions. Rather than looking at formal constraints (such as legal frameworks) as stable, unchanging entities, it regarded these as "sedimented products of contingent beliefs and preferences" (p. 41). Translated to my research, this meant that the paths or institutions limiting the possibilities for constitutional reform were not so much found in, for example, the constitutional clauses formally overseeing that procedure but, rather, in the connected beliefs and habits in which this procedure and the constitution itself was embedded, intentionally or unintentionally. It was to be expected that constitutional lawyers would take an authoritative role in explaining what these beliefs and habits—in other words, traditions—were, even when their competition for interpretative sovereignty could not be fully separated from the political environment in which their discussions were inevitably embedded.

As the next section will explore in greater detail, this somewhat eclectic and loosely built analytical framework—not unusual for a historian—allowed me to better integrate the law into my political history narrative. At the same time, I was not fully aware that the term tradition in particular had led me into a conceptual minefield when it came to the legal discipline itself, specifically legal history. Operating very much in the service of classical legal scholarship in favor of a closed and coherent body of doctrine, traditional legal historians have primarily used the term tradition in the singular to describe a "set of deeply rooted, historically conditioned attitudes about the nature of law, about the role of law in society and the polity, about the proper organization and operation of the legal system, and about the way law is or should be made, applied, studied, perfected and taught" (Merryman & Pérez-Perdomo, 2007, as cited in Duve, 2018, p. 21). Typically, tradition is used here for rather large spatial realms: the common law tradition of the Anglo-Saxon world, the civil law tradition on the European continent, and so on. More recently, however, legal historians, inspired by insights and methods from the humanities, have started to move away from such essentialist approaches. This has resulted in more contingent and practice-oriented understandings of legal tradition, defining tradition, for example, as "normative information that is produced, captured and adapted by communities of practice" (Duve, 2018, p. 30). Although the analysis described in the next section was not informed by these influences, the turn made by legal historians clearly opens roads for an even better integration of disciplines. The conclusion will return to this issue.

In my empirical analysis, however, tradition served in the first place as an instrument to describe the normative dimension of constitutional legal knowledge and beliefs, whether expressed in ideas, habits, or practices. Moreover, having learned that coherence of the law is essential in formalist legal argumentation, I assumed that what was imagined to be constitutional tradition, consequently, also influenced constitutional reforms, as the lawyers helping to shape these reforms would be inclined to prefer proposals that were consistent with tradition over those that were not. To trace Dutch constitutional tradition, I reread the committee reports, legislative proposals, and constitutional legal writings, which I had initially studied for their actual reform plans. This time, I searched for a deeper layer of beliefs, habits, and narratives about what ("good") constitutional law was or should be. Eventually, I identified three concurring traditions connecting a particular view of the past to the shape of future reforms.

## Constitutional Reform in the Postwar Netherlands

Before looking at these traditions in more detail, a few words on the empirical case of the Netherlands that provided the context for this conceptual exploration are necessary. As I will briefly explain, the case of the Netherlands is somewhat exceptional, necessitating the rather broad manner of defining tradition in the section above. As was already highlighted in Donner's 1948 observations, the postwar Netherlands did not particularly feature as the heyday of the authority of written constitutions and constitutional legal scholarship. Ever since, Dutch constitutional lawyers have continued to question the normativity of Dutch constitutional law (Voermans, 2019), while doctrinal debates are observed to be lacking (Zoethout, 1997). This is partly a matter of comparison: Where globally constitutional politics is observed to be firmly on the rise (Hirschl, 2004), the Dutch constitutional system lacks one of the features through which such politics often takes place—a constitutional court. Notwithstanding recent reform attempts, the Dutch constitution thus far has left the last say on the constitutionality of legislation in the hands of the legislator itself. This means that not only the formal reform of the constitution but also its eventual (re)interpretation is decided "in the Hague," notably in the Senate, where legal argumentation tends to blend with more political considerations.

Political considerations also provided the impetus for the debate about constitutional reform that began during the early postwar years. Unlike many neighboring states, the Netherlands had no urgent need to introduce a completely new constitution. Yet, the return to the 1814 constitution, which had undergone major reforms in 1848 and 1917, was accompanied by numerous proposals for fundamental reform of the political system. Once the most urgent reforms—including the decolonization of Indonesia—had been dealt with in the late 1940s, the next decade saw the establishment of a heavyweight committee of political leaders and constitutional experts to prepare an overall modernization of the constitution. Soon, however, the condition that proposals could garner the approval of a broad political majority—the Dutch constitution requires not just two legislative rounds but also a two-third majority in both chambers of parliament for any constitutional reform to be approved—proved insurmountable. The broad political compromises that enabled the development of the Dutch welfare state during those years did not extend to the very foundations of the political system, as many reformers had hoped.

Eventually, modernization of the constitution only took place in 1983, following a legislative operation that stretched out over almost a decade. Two phases separated 1983 from the failed 1950 committee. First, the publication of the *Proeve van een nieuwe grondwet* (1966), a draft constitution prepared by civil servants in dialogue with legal scholars, explored the possibilities and political salience of reducing the constitution to "the very minimum": a technical document that, released from its historical shape and language, contained only the most essential norms regulating the system of government and its limits. Second, another heavyweight constitutional committee of (former) politicians and constitutional experts was established in the late 1960s under the leadership of Jo Cals and André Donner in order to condense the over 60 societal responses provoked by the *Proeve*, together with the new radical plans for political reform debated at the 1967 elections, into coherent and convincing legislative proposals for constitutional revision (Van Leeuwen, 2013).

My reading of the endless minutes of that Cals-Donner committee, which convened seasoned politicians and law professors with a new generation of political scientists in elaborately decorated *Binnenhof* backrooms, provoked me to look beyond the mere arguments the committee members exchanged and to try to understand them not only in a political context but also as part of ideas and narratives inspired by diverging disciplinary traditions (Fig. 1). At times, it seemed as if the members of the committee spoke different languages. This Babylonian confusion most prominently featured in the often-heated debates regarding the radical reform plans tabled at the 1967 elections that addressed the electoral system and the procedure for government formation, among other things. While some constitutional lawyers proposed tackling these issues by removing its regulation from the constitution—very much in line with the minimal constitution proposed in the *Proeve*—political scientists continued pleading to introduce a range of new provisions based on a systematic analysis of the flaws of the reigning conventions (Van Leeuwen et al., 2020, pp. 447–55). The latter's implicit and explicit references to Thorbecke, whose 1848 constitutional reform had actively transformed the 1813 Kingdom of the Netherlands into a constitutional monarchy, hinted at underlying disagreements about what the constitution was meant to do.

**Fig. 1** Diverging ideas and narratives in the 1967–1971 Cals-Donner committee; collage based on a group portrait made by Stokvis to mark the official farewell of the committee March 29, 1971 © Van Leeuwen

## Three Constitutional Traditions

The ideas embodied both in the *Proeve* and by (interpretations of) Thorbecke provided the starting points for two of the three "traditions" I eventually distilled from Dutch constitutional thought and practice. These traditions roughly succeeded each other chronologically in terms of dominance: As new paths worn into soft soil, the latter two steered Dutch constitution-making in the twentieth century "away from Thorbecke's tracks." At the same time, as the debates in the Cals-Donner committee demonstrated, older traditions continued to exist as sets of beliefs and habits influencing what constitutional reforms should look like and how they should be organized. Eventually, the 1983 constitutional reform at best resembled an amalgamate of all three traditions, with different traditions inspiring different parts of the legislative operation, as a closer look at these traditions will illustrate (Van Leeuwen, 2014).

The first of the three traditions influencing twentieth-century constitutional debate referred back to its nineteenth-century origins, more specifically the 1848 constitutional reform. In this Thorbeckean tradition, the constitution featured as an instrument of change (Elzinga, 1998). In 1848, the constitutional reform was meant to pave the way for new political and societal structures, such as the uniform system of local and regional government or a more active education policy. It did so by including the basis for these new legislations in the constitution: After decades of government by royal decree, the new constitutional framework meant to introduce a firm rule of law (*heerschappij der wet*). Hence the belief central as well in later manifestations of this tradition that constitutional revision was foremost an instrument heeding systemic reform—whether it be Thorbecke's ministerial responsibility in 1848, the introduction of general suffrage in 1917, or the new procedure for government formation proposed in the 1960s.

Where the Thorbeckean tradition leaves no other way to such reform than through constitutional revision, the early twentieth century witnessed the emergence of alternative views of the constitution and its ability to accommodate reforms. In this period of growing democratization, the constitutional framework was increasingly experienced as galling bonds (Verkouteren, 1912). In a political landscape of minorities, the formal reform procedure with its demand of a two-third parliamentary majority blocked developments that were widely demanded in political debate or, as legal scholars recounted, were considered legitimate in national legal consciousness. Some issues, such as the pressing political issue of private (in the Dutch context, confessional) schools, were eventually accommodated by parliamentary majorities reinterpreting the relevant constitutional clauses. Constitutional lawyers encouraged this more flexible interpretation as a necessary step toward the demands of an expanding electorate. At the same time, they warned that the constitution was losing credibility and carried the risk of encouraging revolution (Krabbe, 1906).

Therefore, they discussed a new approach to the constitution that left more room to democratic politics by moving away from the "typical, legal interpretation applied by solicitors to a contract, or by judges to a statute" (Van Hamel in Handelingen N.J.V., 1914, p. 166). Instead, the constitution, in the words of constitutional law professor A. A. H. (Teun) Struycken, should be interpreted as a historical national document and a guiding principle for national policy (Struycken, 1914). According to this Struyckean tradition, reform no longer needed to be preceded by constitutional revision: As long as it fit the broader constitutional framework of legitimate decision-making, some stretching of constitutional limits was allowed. Eventually, a formal revision would then follow to again incorporate the main achievements of those democratic reforms in the constitution—if only to make sure that the constitution continued to reflect national legal consciousness and thus guarantee the authority of the constitutional framework. From its Thorbeckean role as pathfinder, the constitution assumed a more passive role. This Struyckean view of the constitution as a primarily symbolic historical document guided many debates over constitutional reform well into the 1950s (and beyond).

A third tradition emerged in the 1960s from the same desire to withdraw the constitution from everyday politics. Yet the constitutional model represented by the *Proeve* at the same time rejected the focus on the symbolic value of the constitution, while reemphasizing its normative, legal value. The *Proeve* proposed a short constitution that would not stand in the way of everyday politics unless fundamental principles were concerned. At the same time, by expanding the catalogue of basic rights, it strengthened the normative safeguards against a still increasing state intervention. Its shortened and modernized text cut the ties with the past, thus seeming to embrace the later often-heard complaint that it was mainly of interest to legal professionals.

## The 1983 Constitutional Revision: Paths Toward (Non-)Reform

When in 1983 the modernization of the Dutch constitution took its final shape, politicians as well as citizens seemed to have long lost interest. The formal announcement of the revision, scheduled on a cold February day, barely made headlines. As the media reported, radical reform plans, such as those proposing a new electoral system or a reform of the procedure of government formation, had already been taken off the agenda in the mid-1970s. Beyond some minor reforms, for example, the introduction of an ombudsman as well as a clause on transparency, lawyers described the 1983 revision as a facelift of the old lady (Heringa & Zwart, 1983): a mostly textual, technical operation that strengthened the catalogue of basic rights, modernized language, and cleared out many provisions considered either outdated or too detailed.

To explain why some reform attempts were successful while others failed, it is important to examine the interests of the dominant political parties. At the same time, I have found that differing beliefs about the constitution also influenced the success of these attempts—beliefs that could be mapped through the various constitutional traditions. Often, the two explanations were narrowly intertwined. For example, as chairs of the constitutional committee dealing with the radical reform plans in the late 1960s, Cals and Donner were, not coincidentally, also members of the confessional parties represented in the confessional-liberal governing majority that did not support these reforms. Yet reformists did not help their case by proposing to radically diverge from what many committee members believed to be the constitutional framework (Van Leeuwen, 2013).

More precisely, the reformists—many of them leftist political scientists arguing from theoretical models about the "ideal" system—proposed introducing pathbreaking changes to the electoral system, among other things. Using the constitution as an instrument of change in a Thorbeckean manner, they argued that "in no other way could practices in government formation be changed" (Glastra van Loon, 1966, p. 135). Meanwhile, many constitutional lawyers instead favored a *Proeve* approach: removing obstacles and, thus, enabling reforms to take shape outside the constitutional framework. In their view, the constitution was "no place for experiments" (Simons in Van Leeuwen et al., 2020, pp. 131–3), and reflecting the Struyckean tradition, they wondered whether the matter had sufficiently "ripened" enough to be integrated into the constitution and were opposed to "declaring politicological conclusions normative and consolidating them in the constitution" (De Pous, respectively Jeukens, as cited in Van Faassen et al., 2010, November 24, 1967). The deep cleavages between the approaches of the two sides even regarded the working order of the committee: Should the committee begin with an open debate about problems and solutions of the political system or with Article 1? With reform plans so clearly diverging from the dominant paths of constitutional revision, it proved impossible to find a satisfactory compromise—both in the committee and in consecutive parliamentary debates.

The constitution eventually promulgated in 1983 did, in fact, mostly follow the *Proeve* path, albeit with some traces of Struyckean thinking. As the lawyers drafting the *Proeve* had suggested, the new constitution drastically reduced the number of provisions and put greater emphasis than before on rule of law elements, for example, with its prominent new chapter on basic rights. Except for some controversial clauses, the text was modernized and shortened. Legal consistency did not prevail in all cases, however. A proposal to subject the newly introduced basic rights chapter to judiciary review was rejected as was an attempt to remove the procedure for appointing local mayors from the constitution (to open that procedure for elections). Moreover, elements such as the new catalogue of social rights, which were explicitly described as having no legal effect other than providing "guiding principles," were reminiscent of the merely symbolic approach to the constitution that dominated the Struyckean tradition.

While these three traditions help to bring out the role of legal thinking and beliefs in the constitutional reform, a better understanding of this legal dimension also enables us to comprehend that the new 1983 constitution, despite a widely felt experience of failure, marked an important transformation after all. This is because the traditions that helped shape the reform also shaped the *use* of the constitution in subsequent years. As one of its drafters soon observed, the new emphasis on the catalogue of rights steered that use in the direction of an increasing focus on its role as safeguarding individual freedoms (Van der Hoeven, 1988). This more normative use of the constitution found its parallel in the emerging jurisprudence based on international human rights treaties that in the Dutch legal system had already been granted direct effect in the 1950s (Van Leeuwen, 2012). Together, these developments suggest that despite the absence of a judicial review of constitutional law, the trend toward constitutionalization with its greater emphasis on rule of law elements and the judiciary also took root in the Netherlands.

## Concluding Remarks

In recent years, research crossing the disciplinary boundaries between law and history has experienced a major upswing. While historians are increasingly aware of the legal elephant in the room when analyzing constitutional, European, or international politics (Patel & Röhl, 2020), lawyers from their side are moving toward more contextual approaches (e.g., Taekema et al., 2020), allowing for a more contingent understanding of the law (Venzke & Heller, 2021). In some legal subfields, that exchange has already produced important conceptual and methodological reflections, such as in the history of international law (Orford, 2021), or in the history of European law (Davies & Rasmussen, 2012). As a concept not limited to the constitutional subfield for which it was developed, the concept of tradition explored in this chapter might help to further this interdisciplinary dialogue.

From a historical, contextual point of view, the concept invites the researcher to approach the law not just as written law, or as the outcome of a political process, but as part of a larger normative framework of ideas, beliefs, and habits usually driven by the aim for consistency—hence precluding radical change. Lawyers may claim an authoritative role in making and explaining the law along the lines of internal coherence. Yet that does not rule out that they compete for interpretative sovereignty among themselves, as well as in dialogue with broader politics or society. In this chapter, the concept of tradition is used to describe those diverging and competing narratives carrying normative information about how constitutional law should be. Elsewhere, I have also tentatively used the concept to analyze a particular trend in Dutch constitutional thinking about and practice vis-à-vis the international legal order (Van Leeuwen, 2012). Here too, the concept helped to highlight how next to other political considerations, contested beliefs about a consistent normative practice informed political decision-making.

At the same time, from a legal perspective, the emphasis this chapter places on tradition in plural form implies that normativity is rendered contingent and subjected to a competition between various narratives. While this loss of coherence may seem to threaten the law's authority, it also opens ways to new coherences. In fact, better appreciating the historical context in which earlier traditions emerged and acquired meaning may inspire us to leave outdated paths and make space to begin imagining new ones.

# References

Andeweg, R. B. (1989). Institutional conservatism in the Netherlands: Proposals for and resistance to change. *West European Politics, 12*(1), 42–60. https://doi.org/10.1080/01402388908424722

Bevir, M., & Rhodes, R. A. W. (2003). *Interpreting British governance*. Routledge.

Bourdieu, P. (1987). The force of law: Toward a sociology of the juridical field. *Hastings Law Journal, 38*(5), 814–853.

Davies, B., & Rasmussen, M. (2012). Towards a new history of European Law. *Contemporary European History, 21*(3), 305–318. https://doi.org/10.1017/S0960777312000215

Donner, A. M. (1948). Grondwetsstudie in Nederland 1848–1948. In J. Valkhoff (Ed.), *Grondwet en maatschappij in Nederland 1848–1948* (pp. 327–362). De Kern.

Duve, T. (2018). Legal traditions: A dialogue between comparative law and comparative legal history. *Comparative Legal History, 6*(1), 15–33. https://doi.org/10.1080/2049677X.2018.1469271

Elzinga, D. J. (1998). De betekenis van de Nederlandse grondwet in de 19e eeuw. In N. C. F. van Sas & H. te Velde (Eds.), *De eeuw van de grondwet. Grondwet en politiek in Nederland, 1798–1917* (pp. 80–95). Kluwer.

Etxabe, J. (2020). Law as politics: Four relations. *Law, Culture and the Humanities, 16*(1), 24–41. https://doi.org/10.1177/1743872116679392

Galligan, D. J., & Versteeg, M. (2013). Theoretical perspectives on the social and political foundations of constitutions. In M. Versteeg & D. J. Galligan (Eds.), *Social and political foundations of constitutions* (pp. 3–48). Cambridge University Press.

Geertz, C. (1983). *Local knowledge: Further essays in interpretive anthropology*. Basic Books.

Glastra van Loon, J. F. (1966). Een Proeve van weinig zeggen. *Acta Politica, 2*, 128–135.

Handelingen N. J. V. (1914). Verslag vergadering, tweede zitting zaterdag 27 juni 1914. *Handelingen Der Nederlandsche Juristen-Vereeniging, 44*(2), 81–177.

Heringa, A. W., & Zwart, T. (1983). Facelift van een oude dame? De grondwet van 1983. *Nederlands Juristenblad, 172*(8), 233–247.

Hirschl, R. (2004). The political origins of the new constitutionalism. *Indiana Journal of Global Legal Studies, 11*(1), 71–108.

Hirschl, R. (2013). The strategic foundations of constitutions. In M. Versteeg & D. J. Galligan (Eds.), *Social and political foundations of constitutions* (pp. 157–181). Cambridge University Press.

Krabbe, H. (1906). De heerschappij der grondwet. *De Gids, 70*(4), 371–407.

Luhmann, N. (1993). *Das Recht der Gesellschaft.* Suhrkamp.

Merryman, J. H., & Pérez-Perdomo, R. (2007). *The civil law tradition: An introduction to the legal systems of Europe and Latin America* (3rd ed.). Stanford University Press.

Orford, A. (2021). *International law and the politics of history.* Cambridge University Press.

Patel, K. K., & Röhl, H. C. (2020). *Transformation durch Recht: Geschichte und Jurisprudenz Europäischer Integration 1985–1992.* Mohr Siebeck.

*Proeve van een nieuwe grondwet.* (1966). Staatsuitgeverij.

Roux, T. (2019). Interdisciplinary synergies in comparative research on constitutional judicial decision-making. *Verfassung in Recht Und Übersee, 52*(4), 413–438. https://doi.org/10.5771/0506-7286-2019-4-413

Struycken, A. A. H. (1914). *De grondwet: Haar karakter en waarde.* Gouda Quint.

Taekema, H. S., Gaakeer, A. M. P., & Loth, M. A. (2020). *Recht in context: Een inleiding tot de rechtswetenschap.* Boom Juridische uitgevers.

te Velde, H. (1997). Politieke cultuur en politieke geschiedenis. *Groniek, 30*(137), 390–401.

van der Hoeven, J. (1988). *De plaats van de grondwet in het constitutionele recht* (2nd ed.). Tjeenk Willink.

van Faassen, M., van Leeuwen, K., & de Valk, H. (2010). *Onderzoeksgids Grondwetscommissies 1883–1983.* http://resources.huygens.knaw.nl/grondwetscommissies/onderzoeksgids

van Leeuwen, K. (2012). On democratic concerns and legal traditions: The Dutch 1953 and 1956 constitutional reforms 'towards' Europe. *Contemporary European History, 21*(3), 357–374. https://doi.org/10.1017/S0960777312000240

van Leeuwen, K. (2013). *Uit het spoor van Thorbecke: Grondwetsherziening en staatsvernieuwing in naoorlogs Nederland*. Boom.

van Leeuwen, K. (2014). Tussen afschaffing en versterking: Historische tradities en de toekomst van een tweehonderdjarige grondwet. *TvCR : Tijdschrift Voor Constitutioneel Recht, 5*(2), 96–116.

van Leeuwen, K., van Faassen, M., & Scherer, M. (2020). *Over de Grondwet gesproken: Een selectie uit de documenten van staatscommissies voor grondwetsherziening 1883–1983*. Uitgeverij Verloren.

Venzke, I., & Heller, K. J. (Eds.). (2021). *Contingency in international law: On the possibility of different legal histories*. Oxford University Press.

Verkouteren, H. (1912). *Grondwetsherziening*. Hollandia.

Versteeg, M., & Galligan, D. J. (Eds.). (2013). *Social and political foundations of constitutions*. Cambridge University Press.

Voermans, W. J. M. (2019). *Het verhaal van de grondwet: Zoeken naar wij*. Prometheus.

Zoethout, C. M. (1997). Grondwet en staatsrechtwetenschap in Nederland 1796–1996. In P. W. C. Akkermans (Ed.), *Twee eeuwen grondwetgeving in Nederland: Staatsrechtconferentie 1996* (pp. 75–95). Tjeenk Willink.

# Rethinking Eastern Europe in European Studies: Creating Symmetry Through Interdisciplinarity

Ferenc Laczó

## Introduction

A glance at European Studies curricula in the Netherlands and across Western Europe reveals a marked pattern of underrepresentation when it comes to Eastern Europe, a region that has at times—and rather revealingly—been called 'the other Europe.' As a historian interested primarily in political ideas and political history, this underrepresentation strikes me as rather curious and in need of reconsideration for several reasons. A case for rethinking can certainly be made in terms of *Realgeschichte*. After all, a host of the most cataclysmic and consequential transformations in recent European history have started and often also played out primarily in Eastern Europe—just think of the Sarajevo assassination in 1914 and the outbreak of World War I (WWI), the Nazi German attack on Poland and the start of World War II (WWII), Nazi

F. Laczó (✉)
Department of History, Faculty of Arts & Social Sciences, Maastricht University, Maastricht, The Netherlands
e-mail: f.laczo@maastrichtuniversity.nl

© The Author(s) 2023
K. Bijsterveld, A. Swinnen (eds.), *Interdisciplinarity in the Scholarly Life Cycle*,
https://doi.org/10.1007/978-3-031-11108-2_5

and Soviet projects of empire building and mass violence, the 1948 coup in Prague and the beginning of the Cold War confrontation, or the *refolutions* in Warsaw and Budapest in 1989 and the end of the continent's East-West division (Garton Ash, 1990). As I shall aim to explain below, beyond such more factual references, the place and role of East European countries also appears critical when we wish to analyse Europe's transformation in modern and contemporary times and address major challenges the European project currently faces.

Starting from the understanding that the notion of 'Eastern Europe' was 'invented' by Western thinkers during the Enlightenment as a space in-between the 'civilized' and the 'barbaric,' this chapter will draw on insights from a host of disciplines—politics, economics, sociology, linguistics, geography, religious studies, and psychology in particular—to try and conceptualize this rather elusive entity with the larger aim of reflecting on East-West dynamics in modern and contemporary Europe. By applying concepts such as 'the demi-Orient' (Wolff, 1994) and 'semi-peripheries in the world system' (Wallerstein, 1974) and, more specifically, theories of 'nesting Orientalism' (Bakić-Hayden, 1995) and 'the East-West slope' (Melegh, 2005) while explicating their multidisciplinary origins, I shall first discuss how and with what consequences Eastern Europe was made to fit into the modern system of integrated and rather homogeneous nation states. I will subsequently try and trace how the East European aspiration to be accepted as 'fully European' has led to the enlargement of the EU after 1989 and resulted in a less well-balanced Union. This part of the chapter will suggest that the just mentioned key aspiration on the part of East Europeans may have directly contributed to the rise of 'illiberal democracies' in countries such as Poland or Hungary—which points to an urgent need to reconsider East-West dynamics and their consequences in a more complex and interdisciplinary manner.

In other words, the current essay draws on and combines insights from a host of humanities and social science disciplines to offer critical reflections on political ideas and develop a new narrative of political history. It shows how the very concept of Eastern Europe has been shaped by various disciplinary discourses and how the asymmetrical relationship between 'Eastern Europe' and 'Europe' can only be properly grasped when thinking in an interdisciplinary manner.

# A Powerful, Malleable Concept

Since the making of modern Eastern Europe is practically inseparable from the power of symbolic geographies, let me begin with some general remarks concerning the *conceptual history* of Europe and the implications of the imaginary Eastern border of this quasi-continent. Since the beginnings of modern history some five centuries ago, inhabitants of Europe have tended to view their continent as substantially different from other ones and have often depicted it as especially valuable, if not downright superior. In the early twenty-first century, Europeans increasingly recognize their continent to be a rather smallish part of the planet that—though undoubtedly influential in certain areas—is no longer of central import in global affairs (Chakrabarty, 2000).

At the same time, numerous Europeans living today believe the European political and social model—typically understood, if not necessarily in so many words, as the partly transnational governance of post-classical, liberal democratic nation states with an embedded form of capitalism—to be preferable to its major, US American and Chinese alternatives and, thus, of continued universal relevance (Jarausch, 2021). In other words, despite having clearly declined in relative terms and no longer exercising a large share of global power, many Europeans hold on, at least implicitly, to the idea that Europe remains a particularly 'civilized' kind of place that other continents could emulate to their benefit.

Given such peculiar continuities in European thought from colonial times into our post-colonial present, it is worth recalling that few key concepts have in fact been as malleable across the millennia as the idea of Europe. This remarkable idea has crossed many boundaries, transitioning from mythology to geography, and then on to religion and culture, to emerge as an increasingly contested political concept in our age—the properly contextual study of Europe across the centuries thus requires a multidisciplinary approach before more interdisciplinary reflections can be developed. As will be familiar, the name of an abducted and raped princess in Greek mythology came to be employed as an—at first, admittedly, rather vaguely defined—geographic expression to refer to a part of the world different from the actual place of Europa's own origins.

During the Middle Ages, this part of the world, coexisting in an often-conflictual relationship with the so-called Islamic world, acquired marked religious connotations: it came to be associated with the realm of Christendom. Importantly for my purposes, the Middle Ages was also the era in which, due to the great schism of 1054, the distinction between 'Eastern' and 'Western' Christianity first acquired seminal importance; a religious distinction on which the concept of Eastern Europe, to be invented during the philosophical Enlightenment of the eighteenth century, would draw, however imperfectly.

The idea of Europe then got connected in early modern times to notions of a specific civilization with claims to universality. It was racialized at the high point of European imperialism in the late nineteenth century with devastating consequences outside and—with the racial imperialism of German Nazism, above all—soon also inside the continent (Mazower, 2008). In the post-war period of the last century, 'Europe' was launched as a novel political-economic-legal project with an unclear end goal—a project that was at first, for rather obvious political reasons, entirely restricted to one side of the Iron Curtain and which, in fact, derived good parts of its *raison d'être* from opposing 'the East' (Patel, 2020).

Self-celebratory *longue durée* narratives may attempt to trace back the origins of Europe to ancient times, to ancient Greece and Rome in particular, but, as a self-conscious project by that name, Europe can be said to be a quintessentially modern invention. A key paradox of nationalism, identified by Benedict Anderson (1983) as the combination of the relative novelty of national consciousness with the retrospective construction of an extended line of continuity, thus, appears to apply to predominant forms of European self-identity too. As more detailed analyses in conceptual history can reveal, many of the traditions the European project claims to embody today were in fact developed by people for whom 'European identity' was at most marginally important.

# Consequential Imaginary Borders

Zooming in on our specific subject of Eastern Europe within European and global frames, it is essential to first consider the puzzling geography of a continent that is not quite a continent and the dilemmas and tensions that have resulted from the underdefined nature of Europe's borders. While water defines the geographic borders of Europe in the West, the South, and the North, this quasi-continent lacks such clear boundaries in the East. The geographic border between Europe and Asia is usually placed somewhere in the middle of Russia. Irrespective of whether it is meant to be constituted by a river or mountain ranges, such a border between continents is, of course, imaginary—and Istanbul's supposed bridging of its European and Asian sides also convinces more as a symbol than as a 'hard fact' (Maçães, 2018). It is worth adding that there is no natural border either between places in continental north-western Europe like the Netherlands and areas in the Far East, which implies, among other things, that Eastern Europe—unlike the narrower area of the Balkans (Mazower, 2000)—cannot be viewed as anything more than a very loose geographic category.

In strictly geographic terms, Europe may be an imaginary continent, but one whose peculiar manner of invention has had serious cultural and political consequences. Constructed as a quasi-continent, Europe has, apparently, been in almost constant need of delineating its own limits vis-à-vis the East to remain distinct while 'Europeans' have also recurrently pushed eastward to spread further the 'ideas of Europe.' While the latter pursuit does not need to take violent forms, the Europe-Asia divide has in fact rarely been conceived in a detached, let alone symmetrical fashion in recent centuries. As it was enforced as the standard way to perceive the modern world, the distinction between these 'two continents' has tended to be not only about a divide in space but also about a gap in time— about 'the synchronicity of the asynchronous,' to cite Ernst Bloch (1935). We might go as far as to suggest that Europeans have often defined Asia not in a positive sense but rather through how that larger part of their landmass was meant to differ from the European peninsula and, more specifically, what it lacked to be quite like it.

Much of what I have just stated may be said to be common knowledge in our post-colonial present. However, there remains an underdiscussed intra-European layer to the same complex of questions. If Europe has recurrently been defined through distinctions between itself and 'the East,' where does that leave Eastern Europe, the part of the world that combines these two notions in its very name? As Benjamin Schenk has shown in his conceptual history, 'Eastern Europe' has first emerged as a subject of scholarship by geographers in the eighteenth century. The term was then used in philology, modern linguistics as well as history before it would have acquired a primarily political connotation in Cold War-era scholarship—without the Eastern bloc states themselves using the term 'Eastern Europe' in any prominent way to refer to themselves (Schenk, 2017).

While the discourses pursued by the just mentioned disciplines have all constructed 'Eastern Europe,' they have delineated it in overlapping but far from identical ways. To take just two examples from *linguistics* and *religious studies* that allude to some of the complexities involved: Polish is a Slavic language (and thus typically studied in Slavic and East European languages departments), but Polish society is predominantly 'Western' (Roman Catholic) in the Christian-religious sense of the term, whereas Romanian is not a Slavic but a Romance language; however, Romanian society is predominantly 'Eastern Orthodox' in the Christian-religious sense.

Remarkably, people's current understanding of the dividing line between Eastern and Western Europe also appears to be impacted by disciplinary context. This was shown, among others, by an intriguing 2008 experiment conducted in Hungary where most students in a geography class correctly identified that Prague lies further to the west than Vienna but their fellow students in a history class insisted that Vienna belonged to 'the West,' whereas Prague—apparently still in the historical shadow of the Cold War at the time—was located in 'Eastern Europe' (Bolgár & Horváth, 2008).

# The Ambiguous Status of the Demi-Orient

Often depicted as that part of Europe which is not 'fully European,' but rather a sort of 'demi-Orient' within the geography of Europe, a multidisciplinary engagement with the study of Eastern Europe also reveals that such a pattern in the realm of ideas has a remarkable parallel in the findings of *socioeconomic studies* regarding the world system as articulated, perhaps most famously, by Immanuel Wallerstein. Tracing the rise of the modern capitalist economy, Wallerstein has attached great import to the role of semi-peripheries which were closely connected, through unequal relations of exchange, to the developed core areas. When it comes to Eastern Europe, cultural demi-Orientalization and socioeconomic semi-peripheralization appear to have gone hand in hand. Thinking about studies of culture and socioeconomic development simultaneously, that is, in a more directly interdisciplinary manner, in fact reveals semi-peripheralization and demi-Orientalization to be logical correlates. When it comes to mapping this region in a global scheme of things, the two approaches with their different interests and foci present almost mirror images of each other. Crucially for my purposes here, the simultaneous potential to include and exclude Eastern Europe underlies both these mappings, creating an unusually ambivalent situation.

When Eastern Europe acquired the political-economic-geographic shape of the Soviet (or Eastern) bloc, it emerged not only as a major focus of area studies on the other side, that is, 'in the West,' but probably also as the most significant part of the world against which the self-declared European project, launched exclusively in Western Europe, now defined itself. While viewing the Eastern half of this quasi-continent primarily as a political threat after its Sovietization in 1947–1948, a threat famously summed up through the concept of 'totalitarianism,' post-war West European perspectives on Eastern Europe, nonetheless, also preserved elements of an older, condescending-Orientalist manner of thinking.

Ambiguities did not stop there. Just as Western discourses sometimes included East European countries as 'Christian' and 'European' and, therefore, viewed them as deprived of freedom and democracy against their will, they could at other times exclude the very same 'Easterners' from a narrower, more Western-style definition of Europe for a variety of reasons. These include seeing Eastern European nations as not properly 'developed,' as places lacking stable forms of statehood and independent social organizations, or even as peoples especially prone to violence, a crude stereotype especially frequently applied to Southeast Europeans or, more colloquially, the people of the Balkans (see Todorova, 1997). Somewhat schematically put, the politicization of the concept of Eastern Europe during the Cold War meant that anti-communist discourses reconciled to the Cold War division of the continent defined 'the East' out of Europe, whereas anti-communist discourses bent on opposing the *status quo* 'in the East' would highlight its Europeanness (Kundera, 1984).

What the concepts 'Asia' and 'Eastern Europe' thus share is that they were both developed externally to the people they were meant to designate. Due to their primarily negative manner of definition (negative in the sense of providing a definition through what something lacks as well as the implied value judgement behind that perceived absence), both categories possessed only limited potential for self-identification. The relationships these terms reflected, and in some sense brought into being in the first place, have differed from one another in one crucial respect though. That is, that East Europeans, unlike their Asian counterparts, have made repeated and, at times, vocal claims to belong to the very same category of Europeans as 'the Westerners' (or, more narrowly speaking, the West Europeans). They have made such claims while members of the latter group could be prone to distancing themselves from them and would also make occasional attempts to reshape them in their own image—clear signs of an asymmetrical and not particularly well-balanced relationship, to the historical development of which I now turn.

## Polarized Occidentalisms

As a result of this asymmetrical and not particularly well-balanced relationship (which conceptual history can help us grasp), the status of Eastern Europe has remained ambiguous and profoundly uncertain. To overcome such asymmetries, various individuals and groups within Eastern Europe have aspired to 'truly belong' to this prestigious 'continent of the imagination' and aimed to transform their own societies in the image they had of countries further west—often with substantial, if at times poorly informed support from Western actors.

Such attempts to modernize and 'Westernize,' however, have repeatedly generated controversies and could yield powerful political responses in the name of local traditions. As various groups in these societies would be opting, and sometimes oscillating, between admiration and rejection, between hopes and resentments of 'Western models of modernity' (Delanty, 2013), a pattern of internal cultural and political polarization crystallized and has been reproduced across generations. In this respect, the controversy between Westernizers and Slavophiles in nineteenth-century Russia might be viewed as something of an archetypical conflict within the political cultures of modern and contemporary Eastern Europe (Walicki, 1975).

This internal polarization between 'Westernizers,' often called 'imitative liberals' in our age, and authoritarian nationalists, again widely labelled 'populists' today, has in many ways been a *consequence* of Eastern Europe's unequal relationship with Western Europe and the West, more broadly. After all, the key difference between imitative liberalism and authoritarian nationalism is the way they define Eastern Europe's relationship to a West perceived to be more developed and liberal.

An interdisciplinary perspective combining insights from socioeconomic and cultural analysis suggests that East European liberals are intent on reproducing the conditions of the socioeconomic core in the semi-periphery through imitation, whereas authoritarian nationalists emotionally revolt against symbolic practices of demi-Orientalization which only tends to reproduce the perception in the West of Eastern Europe not being 'properly Western' (Krastev & Holmes, 2019). These are two predominant, if contrasting and polarized versions of East European Occidentalism that have come to shape a turbulent recent history full of sudden ruptures and unexpected reversals.

It is also worth noting in this context that there is no comparable internal polarization within West European societies that could be interpreted as the consequence of these societies' relationship to and assessment of phenomena in Eastern Europe. While it is thus important to study the variety of Orientalism in Western Europe when it comes to

Eastern Europe, it is perhaps even more essential to grasp the special variety of Occidentalism in Eastern Europe (Buruma & Margalit, 2004).

## Devastating, Ironic Consequences in Twentieth-Century History

After *a conceptual historical introduction* that drew on insights from a multiplicity of disciplines and combined them to create a new interdisciplinary conceptualization of 'Europe and its East,' and the sketching of a relational approach *to political ideas* focused on two prevalent and polarized versions of Occidentalism in Eastern Europe, let me turn to *political history* more directly. Employing such an interdisciplinary conceptualization and relational approach to political ideas to try and rethink political history means at least two things. It implies that the dynamics of East European political history can only be properly grasped when East-West interactions within Europe and, more specifically, the Western 'models of modernity' that East European actors have constructed and contested are foregrounded. An interdisciplinary approach to how the asymmetrical relationship between the 'two halves' of the continent has played out in political history also allows us to offer original insights into how East-West dynamics have come to define modern and contemporary Europe.

The first major attempt to create a post-imperial Eastern Europe and establish it as part of a broader European and Western political project was made through the 'Versailles system' introduced at the end of WWI. Inspired by the ideas of US President Woodrow Wilson and drawing on French political traditions in particular, the key ambition at the time was to introduce a system of democratic nation states. Significantly enlarged, re-established, or newly created countries such as Romania, Poland, Czechoslovakia, and (what came to be called in 1929 as) Yugoslavia were meant to serve as the key pillars of this system (Connelly, 2020). Nation states with their titular majorities and—often only nominally—protected minorities thus became the dominant form of statehood in what was, at the time, a much more diverse and mixed macro-region of Europe than Western Europe. Even the countries with relatively large titular majorities, such as Poland or Romania, still had around 30% minority populations in the interwar years.

Although the post-imperial system introduced at the end of WWI reflected the aspirations of key political actors and broader mass movements within Eastern Europe to create something akin to what existed further west, the first attempt to remodel this 'demi-Oriental' sphere along Western lines backfired and had disastrous mid-term consequences. The new democratic regimes in the region exhibited numerous weaknesses and shortcomings. Except for Czechoslovakia, they were soon overthrown across Eastern Europe.

What was worse, new totalitarian-imperial orders were soon imposed with extreme violence by Nazi Germany and the Soviet Union, and often also with notable levels of local support and complicity (Snyder, 2010). Minority populations consisting of millions and millions of individuals were meant to enjoy institutional protection on the national and—via the League of Nations—international levels. As a matter of fact, their members were often discriminated with many of them persecuted and eventually murdered, with East European Jews during WWII foremost among them; Eastern Europe being the main centre of Jewish life prior to the unprecedented Nazi and collaborationist onslaught, about 95% of all victims of the Holocaust came from this broad region (Laczó, 2018).

With hindsight, the first major attempt to remodel Eastern Europe along Western lines of democratic nation states thus not only failed to fulfil its promise but the attempt contributed to unleashing cataclysmic processes that had been considered unthinkable within the geography of Europe. Devastatingly and more than a little ironically, what was meant to be a new, post-imperial version of Eastern Europe and part of the 'new Europe' soon became subjected to previously unseen levels and brutality of racial violence. This vortex of violence admittedly started already in the late nineteenth century, taking particularly egregious forms within the disintegrating Ottoman Empire to reach its peak across Eastern Europe during the 1930s and 1940s—a region Timothy Snyder (2010) famously referenced as the bloodlands between Hitler and Stalin. The dramatic experiences of Eastern Europe soon made new concepts invented by scholars from the region to denote mass crimes, such as crimes against humanity and genocide, enter the legal vocabulary and common parlance worldwide (Sands, 2016).

The devastating vortex of violence in Eastern Europe in the first half of the twentieth century and mass immigration into Western Europe in the post-war period combined to assure a great reversal. Before the end of the century, the macro-region of the continent that had entered modernity as a much more diverse one came to be divided into ever smaller and—as a general tendency—more homogeneous nation states. In the post-war decades, the previously rather homogeneous nation states of Western Europe became significantly more diverse (Gatrell, 2019).

The cultural and political consequences of this momentous reversal became plainly visible the latest around 2015—to the incomprehension of many commentators in Western Europe and not only there. Based on their response to the humanitarian crisis related to refugees and migrants, East European political leaders and large segments of local populations appear to have 'normalized' the relative ethnic homogeneity that had been created through so much violence in recent history. While the incomprehension at such rejectionism was understandable, the utter historical novelty of a diverse Western half of Europe and a system of rather homogenous nation states in the Eastern half was rarely noted at the time—which, clearly, has to do with the fact that this perplexing 'great reversal' between Europe's 'two halves' is yet to be studied and discussed more comprehensively.

What should nonetheless be clear is that the first major attempt to remake Eastern Europe in the image of Western Europe via the nation state principle not only meant the temporary triumph and utterly tragic failure of an imitative form of Occidentalism but has also produced a thoroughly ironic result—after all, Eastern Europe today looks a lot more like Western Europe did about a hundred years ago, but that is also true the other way round.

# The 'Europeanization of Eastern Europe' and the Remaking of the EU

The peculiar entanglements and asymmetrical relationship between the 'two halves' of the continent have certainly not disappeared with the second great westernizing revolutions of East European peoples in 1989. Just when many Western intellectual debates revolved around Francis Fukuyama's liberal teleological thesis on 'the end of history' (1992), East European societies were embarking on a highly complex, indeed unprecedented transformation out of the party state and the planned economy.

Disoriented due to the sudden collapse of Soviet rule and the disappearance overnight of their life-worlds, members of these societies generally wished to assert their 'Europeanness' shortly after 1989 and to be perceived as akin to their West European counterparts (Laczó & Lisjak Gabrijelčič, 2020). The signifier 'Europe' arguably alluded to a new utopia of sorts in Eastern Europe undergoing its painful post-communist transformation—it referenced a land of plenty, liberty, and security. At the very same time, West Europeans tended to re-assert their right to measure and assess the 'Europeanness' of East Europeans. In the case of the European Union (a West European Union in all but name at the time), this was done primarily through the 1993 introduction of the Copenhagen criteria according to which they could judge whether countries of the 'other Europe' rightfully belonged to the European project or fell short of 'European standards.'

The horizon of 'EU accession' certainly helped aspiring countries remain relatively stable and stay on the course of liberal transformation, with the post-Yugoslav states constituting a most tragic exception. However, being expected and eager to fulfil a plethora of external conditions in a moment of nominal democratization answered to many of the central questions of national political life before more substantial debates could have taken place. As reflected in Ivan Krastev and Stephen Holmes' famed imitation thesis (2019), the efforts of Eastern Europeans after 1989 always had something akin to the unequal and one-sided way

Jewish assimilation tended to be conceived back in the nineteenth century.

According to Krastev and Holmes' primarily socio-psychological explanation (2019), it was this asymmetrical power relationship and the concomitant imitation imperative in Eastern Europe after 1989 that has yielded acts of rebellion—rebellions of resentment and self-assertion—in more recent years, a massive reversal that should perhaps be viewed as another pendulum swing between two forms of Occidentalism. If the imitative core of the interwar 'Versailles system' in Eastern Europe was replaced after WWII by a rejectionist form of Occidentalism, we may have experienced a comparable, if certainly less radical swing between two opposed forms of Occidentalism since 1989.

The drawn-out process and uneven success of EU enlargement has also introduced new hierarchies between the 'more' and the 'less' Europeanized. Such new hierarchies have fuelled what anthropologist and comparative religion scholar Milica Bakić-Hayden (1995) has called nesting Orientalism. A key insight of Bakić-Hayden's theory is that Orientalism does not necessarily function in a straight-forward fashion with clear sides, that is a distinct West Orientalizing a distinct East. Much rather, Orientalizing practices also take the form of 'exclusionary self-inclusion' in the West. They can and do get encoded into relations between immediate neighbours in a sort of chain, with local Westernizers being especially prone to delineating themselves and their country from those 'further East,' as in the cases of Germans from Poles, Poles from Ukrainians, Ukrainians from Russians, etc. Such practices of nesting Orientalism reinforce what global sociologist Attila Melegh (2005) has termed the East-West slope.

It is indeed conspicuous today how it is precisely those East European countries, such as Poland or Hungary, which had been the first and most eager to 'Westernize' (or 'Europeanize') after 1989, which then ended up electing and re-electing governments with exclusivist visions of Europe and Europeanness. Instead of internalizing the liberal-normative project of the West, the current PiS- and Fidesz-led governments in these two countries have started to propagate religiously connoted, civilizational, even implicitly racist ideas of what Europe ought to stand for.

Their much-discussed illiberal projects may indeed severely damage the rule of law and thus pose an existential challenge to the European Union. However, such a turn to exclusionary illiberalism might in fact be less paradoxical than often assumed. The fact that Eastern Europeans, who have been asserting their right to belong to Europe in recent decades (and whose preparedness to truly belong has been measured via numerous criteria), have increasingly insisted on their right to exclude others from Europe, may also be interpreted as a direct consequence of a certain logic of Europeanization. In accordance with the theories of Bakić-Hayden and Melegh, 'Europeanizing yourself' and 'Orientalizing others' may be approached as two sides of the same coin when it comes to post-communist Eastern Europe; they may be viewed as part and parcel of the same refocusing on Europe that began already prior to 1989 and that, indeed, needs to be analysed in a global frame (see Mark et al., 2019).

There is more to East-West dynamics that is of relevance to understanding contemporary transformations in Europe. As mentioned above, Western discourses during the Cold War sometimes symbolically included Eastern Europe as 'Christian' and 'European' places that were thus deprived of 'freedom' and 'democracy' against their will (Stovall, 2021) and should be considered as potential parts of an 'enlarged historical West' (Sakwa, 2017). Political and cultural actors from Eastern Europe in turn tended to make the supposed connections between Christianity, Europeanness, freedom, and democracy more explicit in the reconfigured Europe after 1989. Government representatives in some of the newer members of the enlarged EU, such as in Poland and Hungary, are certainly among the most vocal today in propagating more exclusionary ideas of Europe and the West—ideas that tend to be considered more controversial, not to say anachronistic further west.

This shows that the second attempt to remodel Eastern Europe along West European lines could also result in certain countries' arrival in a place where the West once was. At the same time, the debates surrounding the illiberal turn of these two nation states modelled on the West at the end of WWI and now parts of an 'enlarged historical West' have recently added to the worsening polarization between two alternative self-understandings: that of the West as a liberal-progressive-normative-optimistic project and a culturalist-racial-nostalgic-resentful one.

However, the entry of East European countries not only impacted the political culture of the newly enlarged West but also contributed to the emergence of a less well-balanced Union. While nearly every second member state of the EU could be called 'post-Eastern' by 2013, in demographic terms the 'newer' ones contained only about one-fifth of the overall Union population. There being significant economic disparities between the western and the eastern 'halves' in the early twenty-first century, the economic share of the latter has in fact remained well below one-fifth (on the history of these disparities, see Janos, 2000). In other words, an economically rather underdeveloped part of the continent containing numerous mostly smallish nation states—which, as we have seen, had been modelled on the West—would come to play a disproportionately large political role on the European level post-2004 in the sense of being responsible for nearly every second Council vote and European Commissioner. At the same time, these 'post-Eastern' member states would come do so without their citizens acquiring anywhere near proportional representation within the EU's own elite (Drounau, 2021). In other words, East Europeans have acquired disproportionate power within the Union via the nation state principle but continue to exert negligible influence via the transnational logic. The main conclusion should be clear: the enlarged EU has been notably less well balanced across various realms and in terms of the composition of its political elites than the Union of 12 founded at Maastricht or its Cold War-era precursors.

## Concluding Remarks

This chapter has drawn on and combined insights from a range of disciplines—politics, economics, sociology, linguistics, geography, religious studies, and psychology—to explore East-West dynamics in modern and contemporary Europe. Following a conceptual historical introduction offering reflections on the ideas of 'Europe' and 'Eastern Europe' by drawing on theories of the 'East-West slope' and 'nesting Orientalism' in particular, my two main aims have been to sketch the variety of Occidentalism in Eastern Europe and to suggest an alternative interpretation of the modern political history of European nation states as well as the contemporary EU.

In other words, through an interdisciplinary treatment of key concepts and complex political ideas, I have aimed to develop a new narrative that foregrounds East-West dynamics to reinterpret European political history. Without my engagement with studies of the modern world system as developed in sociology and economic history, post-colonial critiques of Orientalism first articulated within literary and cultural studies, and analyses of symbolic hierarchies and how they shape aspirations as presented by anthropologists, among representatives of other disciplines, I would not have been able to rethink the asymmetrical relationship and the historical process I was interested in.

Such an interdisciplinary approach to the example of 'Europe and its East' was not only meant to show the benefits of thinking explicitly about how various disciplines of the humanities and the social sciences have come to shape complex concepts and what we can gain by combining their insights. This approach also points to a broader and urgent need to think about intermediate, ambivalent statuses in the global system and their implication for global history. As such an exercise regarding the problem of Eastern Europe shows, inclusion and exclusion may be far from pure categories easy to dichotomize while their dialectic can still have momentous consequences.

More concretely, I have aimed to show throughout this chapter how both a broadly interdisciplinary and properly historical approach to the European project of the early twenty-first century can help us grasp how discursive polarization and institutional misbalances have become intertwined again, giving rise to a renewed sense of an East-West divide. By considering such avenues through which East European positionality and experiences can be included in discussions of Europe in a more substantial and critical manner, this chapter has ultimately intended to provide new impulses to the broader interdisciplinary field of European Studies.

# References

Anderson, B. (1983). *Imagined communities: Reflections on the origin and spread of nationalism*. Verso.

Bakić-Hayden, M. (1995). Nesting orientalisms: The case of Yugoslavia. *Slavic Review, 54*(4), 917–931.

Bloch, E. (1935). *Erbschaft dieser Zeit*. Oprecht & Helbling.

Bolgár, D., & Horváth, G. Y. C. S. (2008). Már nem Kelet, még nem Nyugat. *Élet és Irodalom, 52*(13), 10.

Buruma, I., & Margalit, A. (2004). *Occidentalism: The West in the eyes of its enemies*. Atlantic Books.

Chakrabarty, D. (2000). *Provincializing Europe: Postcolonial thought and historical difference*. Princeton University Press.

Connelly, J. (2020). *From peoples into nations: A history of Eastern Europe*. Princeton University Press.

Delanty, G. (2013). *Formations of European modernity*. Palgrave Macmillan.

Drounau, L. (2021, January 24). Geographical representation in EU Leadership Observatory 2021 of European Democracy Consulting. *EU Democracy Consulting*. https://eudemocracy.eu/geographical-representation-eu-leadership-observatory

Fukuyama, F. (1992). *The end of history and the last man*. The Free Press.

Garton Ash, T. (1990). *The magic lantern: The revolution of '89 witnessed in Warsaw, Budapest, Berlin, and Prague*. Granta Books.

Gatrell, P. (2019). *The unsettling of Europe: How migration reshaped a continent*. Basic Books.

Janos, A. (2000). *East Central Europe in the modern world: The politics of the borderlands from pre- to postcommunism*. Stanford University Press.

Jarausch, K. (2021). *Embattled Europe: A progressive alternative*. Princeton University Press.

Krastev, I., & Holmes, S. (2019). *The light that failed: A reckoning*. Penguin.

Kundera, M. (1984, April 26). The tragedy of Central Europe. *New York Review of Books, 31*(7). https://www.nybooks.com/articles/1984/04/26/the-tragedy-of-central-europe

Laczó, F. (2018, January 29). The Europeanization of Holocaust remembrance. *Eurozine*. https://www.eurozine.com/the-europeanization-of-holocaust-remembrance

Laczó, F., & Lisjak Gabrijelčič, L. (Eds.). (2020). *The legacy of division: East and West after 1989*. CEU Press-Eurozine.

Maçães, B. (2018). *The dawn of Eurasia: On the trail of the new world order*. Penguin.

Mark, J., Iacob, B., Rupprecht, T., & Spaskovska, L. (2019). *1989: A global history of Eastern Europe*. Cambridge University Press.

Mazower, M. (2000). *The Balkans*. Penguin.

Mazower, M. (2008). *Hitler's empire: How the Nazis ruled Europe*. Penguin.

Melegh, A. (2005). *On the East-West slope: Globalization, nationalism, racism and discourses on Central and Eastern Europe*. CEU Press.

Patel, K. K. (2020). *Project Europe. A history*. Cambridge University Press.

Sakwa, R. (2017). *Russia against the rest: The post-Cold War crisis of world order*. Cambridge University Press.

Sands, P. (2016). *East West Street: On the origins of "genocide" and "crimes against humanity"*. Alfred A. Knopf.

Schenk, B. (2017). Eastern Europe. In D. Mishkova & B. Trencsényi (Eds.), *European regions and boundaries: A conceptual history* (pp. 188–209). Berghahn.

Snyder, T. (2010). *Bloodlands: Europe between Hitler and Stalin*. Basic Books.

Stovall, T. (2021). *White freedom: The racial history of an idea*. Princeton University Press.

Todorova, M. (1997). *Imagining the Balkans*. Oxford University Press.

Walicki, A. (1975). *The Slavophile controversy: History of a conservative utopia in nineteenth-century Russian thought*. Oxford University Press.

Wallerstein, I. (1974). *The modern world-system I: Capitalist agriculture and the origins of the European world-economy in the sixteenth century*. University of California Press.

Wolff, L. (1994). *Inventing Eastern Europe: The map of civilization on the mind of the enlightenment*. Stanford University Press.

# Gift and Reciprocity in the Aftermath of the 2003 Heatwave: Using Social Theory to Understand Public Confusion in Response to Solidarity Day in France

Paul Stephenson

## Introduction

The year 2003 was the hottest summer on record, with an additional 15,000 extra deaths recorded in France. More than 80% of excess deaths were in people over 75 years old (Grynszpan, 2003, p. 1169). Analyses of the disaster pointed to a strong link between solitude, social exclusion, and death. Toward the end of a fortnight of excessive August temperatures—two-thirds of France's weather stations reported temperatures in excess of 35 degrees centigrade, while 15 percent registered 40 degrees—as the death toll became apparent, the media reported a breakdown in the country's social cohesion and a failure to protect the most vulnerable in society.

On 26 August 2003, in the immediate aftermath of the heatwave, Prime Minister Jean-Pierre Raffarin personally interpreted the human

P. Stephenson (✉)
Department of Political Science, Faculty of Arts & Social Sciences,
Maastricht University, Maastricht, The Netherlands
e-mail: p.stephenson@maastrichtuniversity.nl

© The Author(s) 2023
K. Bijsterveld, A. Swinnen (eds.), *Interdisciplinarity in the Scholarly Life Cycle*,
https://doi.org/10.1007/978-3-031-11108-2_6

catastrophe as symptomatic of a lack of solidarity due to a weakening of the "social fabric" (Ogg, 2005, p. 14). Employing words such as "together," "we fight," and "our country," he called for strengthened solidarity and for the task ahead to be shared between government and citizens; an explicit display of solidarity would need to be organized. He proposed abolishing a public holiday to finance a new pillar of social security aimed at financing retirement homes and care for people who live with disabilities. Employees would work an extra day (in this case a public holiday) but without extra earnings, while employers would contribute employee costs to a special fund. The rationale was that the most active would help the most vulnerable—a "social contract" of one's day's duration, imposed by the state. It would be a politically orchestrated extravaganza, similar perhaps to a charity rock concert or telethon. While this idea seemed like a knee-jerk response, it was likely a well-thought-out political initiative—solutions go looking for problems (Kingdon, 1995). Such schemes had been tried in Sweden and in Germany, post-reunification with a "solidarity tax." The subsequent law of 30 June 2004 gave a legal framework to raise 2 billion Euros, equivalent to one-fifth of the tax credits the state accords in its budget for older people. The money would equip residential homes and provide better risk management, with 800 million Euros allocated for the people who live with disabilities and 1.2 billion Euros for older people (or rather, the state institutions caring for them).

Despite a huge communications campaign budgeted at 3 million Euros, the day was chaotic: some workers stayed at home assuming it a national holiday; some took the day off officially by using an RTT (compensation day for excess time worked; *réduction du temps de travail*); others worked or at least tried to go in. There were major abstentions in schools and public transportation, but the French government claimed to be satisfied, nonetheless. After three years of confused Solidarity Days, in December 2007, the secretary of state made responsible for evaluating public policy six months earlier, Eric Besson, submitted a report to Prime Minister François Fillon. Deeming the measure "a real success," he nonetheless proposed three scenarios: firstly, sticking with the idea of an obligatory day of work but shifting it to a new date; secondly, reinstating the public holiday and leaving it open to firms to decide upon their day; or thirdly, keeping "Pentecost Monday" but improving childcare on the day.

In February 2008, the minister of work, social relations, and solidarity, Xavier Bertrand, called instead for a day of solidarity *à la carte*, be it an RTT, two half RTTs, or seven hours throughout the year. Prime Minister Jean-Pierre Raffarin, though no longer in office, later sought to resurrect the day. In July 2011, the Constitutional Court upheld the legal basis for the day, finding that it was in respect of the values of the Republic, despite France's biggest union, the CFDT (French Democratic Confederation of Labour; *Confédération française démocratique du travail*) having argued that it undermined the notion of equality, since retired persons and the self-employed were excluded.

## How to Understand the Case?

I was working in France, managing Interreg projects financed by the EU Structural Funds, which aimed at promoting translational cooperation on cross-border sustainable development programs. As such, I wasn't actively pursuing academic research at the time but would subsequently move to Maastricht to take up a teaching position and join the department of history, and then, political science. My postgraduate training had not really provided me with much training in historical or social and political sciences methods; much I would later learn on the job. Neither did I have a background in sociology or philosophy; my higher education had principally been in modern languages (French and Spanish) with some European integration. Within my French studies, I had taken modules on French management and modern politics, so I had a little background understanding of the French state.

Nonetheless, I was fascinated by the political fallout from the 2003 heatwave, and living in France from 2002, as well as previously studying there, I was conscious of a number of structural and cultural factors that might have contributed to the high death toll, not least the almost sacred nature of the summer vacation in France when the country effectively shuts down. In France, you are either a *juilletiste* or *aoûtien*, depending on which month each year you take your holidays. I had witnessed first-hand how large cities emptied out in the summer heat with much of the population (including the political class) at the coast. It meant that those without family, on low incomes, or too frail to travel were left behind, often high up in apartment buildings without elevators.

I collected a lot of material at the time, from magazines and websites, including political cartoons, but it was only once back in academia that I set about trying to seriously analyze the heatwave and its aftermath. My first article (Stephenson, 2009) examined some of the lessons from the heatwave. Political mismanagement had clearly contributed to the death toll with government initially blaming medical services. However, other politico-cultural, societal, and psychological factors have contributed to the failure to protect the most vulnerable citizens. I identified 20 obstacles ("pathogens") to ensuring effective response in the face of environmental or weather-related threats, distinguishing between state-institutional and individual-community barriers, most of which have a cultural dimension. These factors require greater consideration by policy makers to improve preparedness for environmental threats in the EU. The disaster raised questions about crisis management and how best to reduce risk for older populations, illustrating the limits of the state in offering social protection through institutionalized solidarity mechanisms, and recognizes calls to strengthen community-capacity.

While my first article sought to trace what happened (or didn't) during the heatwave—focusing largely on political mismanagement and lack of preparedness—my second and third articles (Stephenson, 2013, 2014) sought to explore what happened in the months that followed. In the 2013 article, I looked at the way in which the heatwave was framed in political debate, including inside the French parliament. It examined the impact of the public health crisis on French public management, considering how government actors across various state institutions, including central and decentralized tiers of public administration, engaged in reform. It studied how these actors in the post-crisis reform process established responsibility and drew lessons. It showed that "solidarity" was used discursively in a game of political blame-shifting and experimentation and it pointed to the politics behind the framing of crisis enquiries (Stephenson, 2013). As such, my main conceptual approach to analyzing the case was to look at the discourse within parliamentary debates and to examine speeches by key actors in politics and public health.

What fascinated me most, however, was a fund-raising initiative advocated by the prime minister—a bold political move to cancel a public holiday and, thereafter, the resistance and refusal, chaos, and confusion that ensued in France, arguably the country with the most generous welfare state and healthcare systems in the world. One couldn't help but be struck by how politicians, including Raffarin, who had been the target of so many accusations of political failure at the time of the heatwave, were now suggesting that the whole country take remedial action to correct previous misdoings—the media had spoken at the time of a lack of intergenerational solidarity. How could one really understand the boldness of this political move? Through what theoretical or conceptual lens might one hope to understand it? How could one conceive of the possible normative justifications for this public policy initiative? One way to approach these questions was to consider the gift-like character of the Solidarity Day.

I was not sure that any of the conceptual tools from political science or public administration could really help me analyze "justifications"; most EU policy analysis I had engaged in thus far was from the perspective of the policy-making cycle and was about doing the detective work to arrive at an evidence-based argument that sought to explain policy outcomes, whether this concerned agenda-setting (who pushed the issue onto the political or media agenda?), decision-making (how can we understand decisions as the result of deliberation and power play?), or implementation (how can we understand sub-optimal policy delivery?). In short, most of my work looked at the role of actors and institutions in advancing their own preferences to drive policy integration. It was interested in identifying the interests and agendas of actors and analyzing how they were advanced.

I could arguably have taken a more systematic approach to analyzing the discourse around the proposal for, and implementation of, the Solidarity Day—perhaps using framing theory (Rein & Schön, 1996) or discursive institutionalism (Schmidt, 2008). Recent work in public administration on experimentalist governance (Sabel & Zeitlin, 2010) conceives of the role of experimentation, deliberation, and informalism as means to social stakeholders and secure consensus on courses of political action. However, I wasn't really seeking to understand policy implementation, or policy failure in practice, but rather, why there was so much confusion around the event and such diversity in the way different groups and individuals reacted. By extension, I was keen to consider, with regard to the role of the state and political philosophy, how one might possibly justify such a bold (and risky) political experiment. And though Raffarin and his government didn't necessarily articulate their rationale in such explicit terms, I was intrigued by how one might—from a historical and societal perspective—justify the move?

Despite no real background in socially or anthropology, perhaps a theory that would help explore the relationships between the state, the welfare state, and citizens would help me contemplate the case, to find compelling arguments. The French sociologist Marcel Mauss' gift theory and public policy together were able to inform my analysis of the response to the introduction of this day, though Mauss' original work did not put forward a method for operationalizing the gift cycle for public policy analysis.

In the process of formulating my arguments, I "pitched" the story of what happened in France as a moral conundrum to a number of friends and colleagues with different disciplinary backgrounds, including philosophy, social anthropology, and sociology, some of whom were familiar with the work of Mauss. By exploring Mauss' ideas and seeking to apply his notions to my contemporary case, I was not consciously engaging in interdisciplinary research, but rather embracing a concept that could help me narrate, discuss, and explore actions and motivations by testing the empirical reality against these notions of gift and exchange at a collective level. I sought and succeeded to place my article in a country-oriented journal, *Modern and Contemporary France*, which is by its very nature a multidisciplinary—though not necessarily interdisciplinary—journal (Stephenson, 2014).

The experience has also made me critically reflect on my own "discipline" of European Studies, arguably really. What constitutes European Studies differs from university to university, but in Maastricht it is heavily oriented to using political science methods, even if, in the broader sense, it is de facto an interdisciplinary "field" that captures political science, history, modern languages, literature, and cultural studies. It is always good to make the distinction between European Studies.

# Mauss and the Notions of Gift and Reciprocity

In 1950, Marcel Mauss wrote of a re-emergence of *gift* as well as the triangle of charity, social service, and solidarity, stating that our lives were permeated with the atmosphere of gift, where obligation and liberty intermingle (Mauss, 1954/1950, 1954/2006). Charity was seen as *a free gift*—"a voluntary, unrequited surrender of resources" (Douglas, 1990/2002, p. ix). Liberty was the absence of all impediments to action, other than physical or mental constraint. The term "free-will" did not infer liberty of will, desire, or inclination, only the liberty of man (Peters, 1956, pp. 167–8). Recently, Aafke Komter (2005) has sought to rethink social ties by bringing together sociological theory on solidarity and anthropological theory on gift exchange, positing that modern theories of solidarity should incorporate core insights from gift theory.

We might consider "gift" to be part of a total system of reciprocity in which the honor of the giver and recipient are engaged. The system is simple: every gift must be returned in some specific way, setting up a perpetual cycle of exchanges within and between generations. In some cases, the specified return is of equal value, producing a stable system of statuses. In others, it must exceed the value of the earlier gift, producing an escalating honor contest. The whole society is thus a catalogue of transfers that map obligations between members, a living record of the credit and debt structure of a community. *Reciprocity* may be seen as the building block of community because it "makes and perpetuates dyadic relationships that are the irreducible core of society" (Gudeman, 2001, p. 80). *Solidarity* is implicit in the gift:

> the state and its subordinate grouping desire to look after the individual. Society is seeking to rediscover a cellular structure for itself. It is indeed wanting to look after the individual. Yet the mental state in which it does so is one in which are curiously intermingled a perception of the rights of the individual and other, purer sentiments: charity, social service, and solidarity. The themes of gift, of the freedom and the obligation inherent in the gift, of generosity and self-interest that are linked in giving, are reappearing in French society, as a dominant motif too long forgotten. (Mauss, 1954/2006, p. 87)

In political philosophy too, *reciprocity* is one of the key concepts, especially when it comes to thinking about what justice (social and distributive) requires: many theories of (re)distribution are based on the understanding of justice as a certain type of social reciprocity between individuals in society. Enacting reciprocity is a tactical act and "way of groping with uncertainty at the limits of a community: offering a gift defends, secures and expands the borders of community" (Mauss, 1954/2006, p. 87). The theory of gift thus is a theory of human solidarity. A gift should enhance solidarity; a gift that does not is a contradiction.

What is *solidaire* about the gift is precisely the in-built expectation of reciprocity, contrary to *free gift* or donation, whereby the donor seeks exemption from any further transaction with the recipient. The gift is thus rarely *free* in day-to-day life but tied up with notions of self-interest and disinterest (Caillé, 2005). Hillel Steiner's (2010) theory of rights and freedom considers interactions between individuals in society as a mere series of transactions: he thus conceptualizes freedom as something that can be measured and located on a range within a continuum. Within this framework gifting (giving something that belongs to you freely away to someone) occupies one end of the continuum; exploitation, in contrast, stands at the other. But what do we mean when we talk of solidarity? Emile Durkheim (1984) distinguished between two types: *mechanical* and *organic* solidarity. In the case of mechanical solidarity in traditional societies, cohesion and integration come from the homogeneity of individuals whereby people feel connected through similar work, education, religion, and lifestyle (Giddens, 1971).

## Applying Mauss to the French Case

In my 2014 article, I set out that Mauss himself would arguably have been fervently opposed to the Solidarity Day, establishing that, as more or less an anti-state anarchist, he would have objected to any form of enforced solidarity, be it via legal and financial channels of the state welfare system. For each hypothesis about gift/reciprocity, there are also constraints in the cycle, that is, there is possibility of pure gift exchange being interfered with or inherently limited due to different factors. We may conceive of multiple motives and competing actor perspectives, when understanding why different parties behaved as they did. How could I better understand the confusion in, and diversity of, public reactions to the proposed Solidarity Day?

First, considering state "gift-giving," I argued that the French government had sought an act of reciprocation for the generous welfare state. The state was merely seeking reciprocity from the citizenry for its many years of massive gift-giving in terms of welfare protection, having put in place a financial and institutional system of social security mechanisms. Arguably, it was now time for citizens to give explicit recognition of the extensive mechanisms put in place to guarantee their welfare and, in so doing, to acquiesce to the state's agenda by participating in the fundraising event. The state was "calling in" a small favor for services rendered, namely to work a day and with no financial consequence—no more and no less money—to help it raise extra tax revenues from firms. The critical problem here was that the state did not seek reciprocation uniformly from all men and women; instead, the tactic was to engage the Republic in an expression of collective solidarity (to raise money *and* strengthen the social fabric). Yet, the strategy promoted the expression of solidarity through work, enabling those with a job to take part but leaving the unemployed marginalized. It promoted the notion of "active citizenship," while reinforcing the idea of dependency by older people on the state. Those gainfully employed could formally "express the Republic" while those unable to work were not presented with alternative means to "give." Paradoxically, social solidarity expressed through everyday support, such as informal and unrecognized social processes, perhaps by volunteers, and which narrows the inter-generational divides, was not recognized. Such day-to-day activity had no visibility or obvious financial contribution to the state. The state's pursuit of a display of *mechanical solidarity* promoted inequality and placed individual liberty at risk.

Second, I put forward that the notion of working an extra day was a gift to older people and, as such, an act of reciprocity also, in recognition of the freedom and equality secured through wartime courage and the post-War social struggle of the 1940–1950s. A typical victim of the heatwave, at 80 years old, would have been born in 1923 and aged 16 at the outbreak of World War II. One might consider the economically active French worker had a political (and economic) obligation to express solidarity (*fraternité*) toward the older people in gratitude for their fighting and resistance which had ensured the worker's own freedom (*liberté*). Moreover, the older generation's political and social struggles in the post-War period had secured greater social, racial, and gender equality, such as with the uprising of 1968. Republican values would have little meaning had the Fifth Republic failed to materialize, for example, if German occupation had endured long-term. The workforce had a political obligation to honor the 15,000 dead, as well as older people, many of whom had been soldiers. Decades later, this was arguably an opportunity to take the gift cycle forward by expressing inter-generational solidarity through work. Workers could exert their free-will precisely because of this gift of brave resilience. At its most fundamental, liberating the Republic had guaranteed the possibility for solidarity to be expressible in 2005.

However, an obvious critique was that older people were not engaged with workers, so extending the gift cycle created impossible burdens. Gift triggers reciprocity and, thereafter, a domino effect of reciprocation, but the theory assumes the ability to reciprocate. Pentecost Monday effectively established a fresh transaction between workers and older people (both dead and alive) who had themselves not sought the intervention of the workforce; the state used them as a scapegoat to coax the economically active into work. Newspaper cartoons I collected at the time featured caricatures of older people, bemoaning ironically, "The young are really going to love us now" or "That was the only day of the year that my son ever visited" (Stephenson, 2009). Older people were effectively newly indebted to the workforce since the healthcare guarantees they had the right to expect from the state were not in place—the financial delivery of welfare was made subject to the actions of the social partners by calling on the Republic. Older people would have to "shoulder" the friction engendered by the possibly disgruntled working population.

Third, I explored the notion that because civil society failed to express solidarity during the heatwave, the state legitimately stepped in to trigger the cycle of gift-giving through forced employment. The supposed breakdown of the "social fabric" in 2003 presented the state with the task of encouraging citizens or "social partners" to take action. Yet, this failure or absence was merely presumed based on a hefty death toll and before any parliamentary enquiry had been conducted and evidence gathered. If indeed political indifference, societal fragmentation, economic imperative, and/or a lack of awareness had prevented appropriate action during the heatwave, then this could now be righted by the state acting as a catalyst for the forwarding of the gift cycle. Moreover, it is the role of the state's managers (elected politicians) to uphold and promote Republican values. This would keep older people high up the political agenda and in the public consciousness. The state would need to act quickly since the public memory changes and fades.

From a different angle, we might also question if the event itself was a "taking back of gift"—here valuable leisure time—on the premise that citizens had chosen inaction over solidarity in August 2003. Pentecost Monday was thus a punishment, made possible by an assumption that the death toll resided with the failure of the citizen, even if this derived from nothing other than media reporting and political rhetoric. The constraint in this case, however, as regards the application of the gift principle, is that workers could not express gift based on unhindered political obligation or free-will. Political coercion to ensure another's security constrains one's own free-will. Solidarity expressed through action must come from an individual's moral choice and *organic solidarity*, not coercion by a third party. In the pure gift-reciprocity cycle giving indeed *becomes* obligation, and by extension, in the political community, membership *demands* obligation.

In short, if the state's ultimate goal was to strengthen community, then orchestrating a collective act of solidarity seemed a logical strategy, even if organized to raise money for public expenditure. As Stephen Gudeman (2001, p. 81) asserts, if a gift is freely given, it has no social impact, because obligations are not set into motion. It is precisely the constraint, coercion, obligation—some derived automatically from membership to the community and some coming from purposeful state action—that was meant to encourage citizen action through experimentation with a balanced form of "economic reciprocity," expressed through services and performances (work). The fund-raising event placed obligations on older people and workers, capturing them within a transactionary cycle. The gift was meant for society at large, the spirit of the gesture had to both strengthen community and act as an expression of *the idea of community*. We might be tempted to consider the collective action as a form of "solidarity" by consensus. However, it could also be argued that the day *could not be* solidarity precisely because the prompt came through a form of legislative state act, tantamount to coercion. It seemed that there were constraints in the pure application of both gift and reciprocity, as well as with the subjective nature of what constitutes gift, charity, and obligation. The case highlights the difficulty in *repaying gift* and the impact of time, between a critical moment of drama and the opportunity to give or reciprocate, on the collective being willing and able to do so.

The first Solidarity Day on Monday, 16 May 2005, and those that followed were a space for political struggle and contestation. The explicit fund-raising initiative was the very visible hand of the state orchestrating and constraining action in the name of older people. This would have appealed to Hobbes' absolutist, supremely powerful state but arguably have repelled Mauss, who, nonetheless, would have been intrigued how the forces of reciprocity, morality, and obligation came into play. He would probably have believed that any expression of solidarity should be free, organic, and of its own accord, without legal prompting, accepting that freedom and obligation, generosity, and self-interest are all somehow inherent in the gift. As the case showed me—as I tested the application of the gift cycle and gift economy in an exploratory mode—this concept developed in the mid-twentieth century still served as a valuable tool for exploring state-citizen relations today.

## Conclusion

Looking back at my analysis using Mauss, I am pleased that I took the leap of faith and engaged with ideas emanating from a discipline beyond my own. That said, because I was early in my academic career and had not received a particularly strong grounding in political science research methods, I can't say that I was particularly conscious of either stepping outside of "my discipline" or actively engaging in interdisciplinary research. Instead, it felt both liberating and fun to engage with the ideas of a thinker whose work I had not read; and, perhaps, it was the very fact that Mauss was not somebody I was supposed to know, that made the process of engaging with his concept of gift and reciprocity so intellectually appealing. Because I first came to the gift cycle from its use in social anthropology in "exotic" cases of places and people far away, there was something challenging—but at the same time logical—about trying to "reign in" the theory and apply to a modern Western European state and the very country he was from.

There have been more recent heatwaves, floods, and freak weather incidents with drastic impacts on some sections of society and where the political, economic, and societal responses could be analyzed from a similar perspective where governments have taken bold crisis management measures. Moreover, Mauss' theory has potential application for analyzing more recent cases such as Brexit and the Covid-19 pandemic. In both cases, there are tremendous questions with regard to inter-generational solidarity. Regarding Brexit, we say a striking contrast in the voting behavior of older versus younger people. Many older, affluent people voted for Brexit, depriving younger people of their EU citizenship and right of free movement, rights that the younger population would arguably make more use of than those whose career is behind them. Older people have enjoyed those rights since 1992. On the one hand, one might argue that the perceived sense of a lack of generational solidarity led the elderly to vote largely for Brexit. On the other hand, one might consider their vote itself as the manifestation of a lack of solidarity. Should votes have been weighted to give more currency to the votes of the economically active, that is, those studying and working for whom the loss of free movement would have greater significance?

In the case of Covid-19, this especially concerns the priority placed in many Western states in 2021 on vaccinating older people over younger people but, in so doing, depriving those who are economically and physically fit of the possibility to travel over the summer. The implications of the pandemic on medical care, economy, and social cohesion raise important questions for social theory about the rights of all citizens, about moral obligation, and, moreover, the limits of the EU's rights of free movement. Likewise, the introduction of the furlough scheme in the UK to pay workers who were unable to work, thus providing them with a secure income in the short-term, completely transformed notions of

Conservative government, the modern state, and social welfare. With massive financial transfers being made to vast swathes of the working population (that regularly pays income tax and national insurance), one could also explore the notion of gift and reciprocity in 2021. Clearly, Mauss' gift cycle still has enormous currency and potential for further application.

# References

Caillé, A. (2005). Don, intérêt et désintéressement: Bourdieu, Mauss, Platon et quelques autres. .

Douglas, M. (2002). Foreword. In M. Mauss (Ed.), *The gift: The form and reason for exchange in archaic societies* (pp. ix–xxiii). Routledge. (Original translation published 1990).

Durkheim, E. (1984). *The division of labor in society*. Palgrave Macmillan. (Original work published 1893).

Giddens, A. (1971). *Capitalism and modern social theory – An analysis of the writings of Marx, Durkheim, and Max Weber*. Cambridge University Press.

Grynszpan, D. (2003). Lessons from the French heatwave – commentary. *The Lancet, 362*(11), 1169–1170. https://doi.org/10.1016/S0140-6736(03)14555-2

Gudeman, S. (2001). *The anthropology of economy*. Blackwell.

Kingdon, J. (1995). *Agendas, alternatives and public policies*. HarperCollins.

Komter, A. E. (2005). *Social solidarity and the gift*. Cambridge University Press.

Mauss, M. (1954). Essai sur le don. In M. Mauss (Ed.), *Sociologie et Anthropologie*. Presses Universitaires de France. (Original work published 1950).

Mauss, M. (2006). *The gift: Form and reason for exchange in archaic societies*. Routledge. (Original work published 1954 by Cohen & West).

Ogg, J. (2005). HEATWAVE: implications of the 2003 French heat wave for the social care of older people. *Young Foundation Working Paper 2*. http://youngfoundation.org/wp-content/uploads/2013/04/Heatwave-October-20051.pdf.

Peters, R. (1956). *Hobbes*. Penguin.

Rein, M., & Schön, D. (1996). Frame-critical policy analysis and frame-reflective policy practice. *Knowledge and Policy, 9*, 85–104.

Sabel, C., & Zeitlin, J. (Eds.). (2010). *Experimentalist governance in the European Union: Towards a new architecture*. Oxford University Press.

Schmidt, D. (2008). Discursive institutionalism: The explanatory power of ideas and discourse. *Annual Review of Political Science, 11*(1), 303–326. https://doi.org/10.1146/annurev.polisci.11.060606.135342

Steiner, H. (2010). Podcast. "On Exploitation." http://philosophybites. com/2010/08/hillel-steiner-on-exploitation.html.

Stephenson, P. (2009). Hot under the collar: Lessons from the 2003 heatwave in France and the security implications for coping with environmental threats in the EU. *Journal of Contemporary European Research, 5*(2), 293–311.

Stephenson, P. (2013). Solidarity as political strategy: Post-crisis reform following the French heatwave. *Public Management Review, 15*(3), 402–415. https://doi.org/10.1080/14719037.2013.769850

Stephenson, P. (2014). Gift and reciprocity at work? Using Mauss to explore justifications for the solidarity day in the aftermath of the 2003 heatwave. *Modern and Contemporary France, 22*(3), 343–361. https://doi.org/10.108 0/09639489.2013.865011

# Freeing the Frog in the Well: Borrowing from History to Understand Contemporary Japanese Development Aid to Ethiopia

Elsje Fourie

Just over a decade ago, the Japanese and Ethiopian governments embarked on an ambitious joint program of "lesson-sharing." Ethiopia's Prime Minister, the technocratic and autocratic Meles Zenawi, would meet regularly with experts from Japan's leading government think-tank to understand and emulate how Japan had achieved rapid industrialization. This "high level industrial policy dialogue" would, in time, result in a number of direct interventions, the most notable of these involving the deployment of Japanese—and eventually Ethiopian—consultants to Ethiopian factories and other workplaces. The aim was to teach Ethiopian workers *kaizen*: the productivity methods and mindsets associated with export powerhouses like Toyota in the 1950s and 1960s. A dedicated government-funded Ethiopian Kaizen Institute (EKI) was established in 2011 for this purpose, and Japan's official aid agency agreed for its part to finance the Japanese consultants. Before long, Ethiopia was being touted

E. Fourie (✉)
Department of Society Studies, Faculty of Arts & Social Sciences, Maastricht University, Maastricht, The Netherlands
e-mail: e.fourie@maastrichtuniversity.nl

© The Author(s) 2023
K. Bijsterveld, A. Swinnen (eds.), *Interdisciplinarity in the Scholarly Life Cycle*,
https://doi.org/10.1007/978-3-031-11108-2_7

as a model for the entire continent due to the fierce commitment of its government to kaizen.

When I began researching kaizen's journey to Ethiopia, I expected to engage primarily with theoretical concepts from political science and anthropology. As a scholar who loosely locates herself in the field of development studies, I felt that these somewhat familiar disciplines would lend academic depth to an empirical phenomenon decried by many as almost laughably facile. A common response to learning of my research topic, especially (but not only) among non-experts, was disbelief and even mild mockery. How could either party possibly think this would work? Did this not precisely demonstrate so many of the age-old problems with foreign aid: that technocrats were forever trying to meddle in local settings without taking culture and context into account?

While I understood these concerns, I wanted to dig deeper. To group all foreign interventions for economic development into one huge category, as international development cooperation's critics often tend to do, risks homogenizing a wide variety of approaches. In the past 70 years, Ethiopia has witnessed every foreign development prescription under the sun. At the moment, projects run the gamut from the construction of a mega-dam by Chinese state-owned enterprises, the training of civil society organizations by the EU, and the provision of emergency healthcare in crisis-affected areas by the United States. Can these all really feature the same underlying dynamics, and what does the existence of this particular Japanese intervention say about the state of development cooperation today? In answering these questions, I needed to explore the social, cultural, and political desires that lay behind both the sending and receiving sides of this particular development intervention. Ethiopian kaizen, I suspected, was more than just a discrete technical fix to an economic problem.

This chapter is the story of how my disciplinary locus shifted during the course of this research, the findings of which were published in the interdisciplinary journal *Global Perspectives* (Fourie, 2020). More specifically, it is the story of how historical case studies, and more specifically the concept of low modernism that emerged from their exploration, came to play an indispensable role. As shall be shown, it was a small group of historians who first observed anthropologist James Scott's influential concept of *high modernism* and created from this its counterpart, *low modernism*. Applying this "traveling concept" to the study of contemporary development cooperation allowed me not only to demonstrate historical continuities between past and present but also to further operationalize low modernism and the move toward its cautious and contingent theorization. Accompanying this personal reflection is, then, a plea for these two fields of study—history and development studies—to engage in more and deeper conceptual encounters that play off the relative strengths of each.

# The Position of History in Development Studies

In order to explain how history came to inform my study into kaizen in Ethiopia—and why its inclusion was by no means a given—it is useful here to briefly explain how the boundaries around the field of development studies have emerged and been maintained.

As with other fields appended by the word "studies," development studies is not a single discipline. In 2017, the European Association of Development Research and Training Institutes (EADI) reached the end of a lengthy consultation into the state of the art. Its resulting definition described development studies as

> a multi- and interdisciplinary field of study that seeks to understand social, economic, political, technological, ecological, gender and cultural aspects of societal change at the local, national, regional and global levels, and the interplay between these different levels and the stakeholders involved. (Mönks et al., 2019, p. 3)

Another highly influential academic body, the Development Studies Association (DSA), locates the roots of the field in "anthropology, economics, sociology, politics and geography" but notes that "it may also combine with others such as psychology, law, management, natural science, history, agriculture or engineering" (DSA, n.d.).

In theory, then, the field should hold a substantial place for history—many branches of which, after all, concern themselves with social change. However, the above conceptualizations are extremely broad and must, therefore, be understood in conjunction with the central empirical questions with which the field has tended to concern itself. A quantitative content analysis of four major development studies journals between 2000 and 2015 found a focus on four common themes: foreign aid, poverty reduction, environmental sustainability, and "development challenges" (i.e., the perceived domestic barriers to development in low-income countries) (Madrueño & Tezanos Vázguez, 2018). More specifically still, development has often centered around attempts by "trustees" to effect progress among groups who are in some way deemed incapable of affecting this change themselves—what Gillian Hart (2001) has influentially termed "big D" development.

Despite recent attempts by the United Nations Sustainable Development Goals (SDGs) to broaden this notion to the Global North, the Global South remains the primary target of such efforts. The concepts "Global North" and "Global South," while controversial, remain in common usage due in part to the lack of a perceived alternative. Without taking a normative stance on their suitability, this chapter retains their mainstream usage for descriptive purposes. The latter term, therefore, "denotes regions outside Europe and North America, mostly (though not all) low-income and often politically or culturally marginalized" (Dados & Connell, 2012), while the former term refers to its economic and political inverse.

If development studies can thus be said to analyze intentional interventions into the improvement of a target population's socio-economic position, we can begin to see why the social sciences might dominate. This would not be inherently problematic were it not for two additional developments in the past few decades. Firstly, social scientists studying development interventions have increasingly been split into two tracks, each of which regards the other with suspicion. The first is a policy-oriented arm, which aims essentially to refine the policy toolbox that trustees have at their disposal, is often populated by development economists and informed by econometric methods. As Bruce Currie-Alder puts it, "the field has shifted away from descriptions of historical patterns of broad social change … toward causal explanation that links particular interventions—in policy or technology—to their outcomes at demonstrable scales or specific dimensions of human well-being" (2016, p. 6). This is in keeping with the general distrust of grand narratives that marked the behavioral turn in many social sciences from the 1970s onward. We see it also in what Charles Gore (2000) has termed the rise of "ahistorical performance assessment," namely the tendency to judge particular countries, governments, or other units of analysis along a uniform metric regardless of their individual development trajectories. Examples of this abound in the various development indexes of bodies such as the UN and the World Bank. This move toward predictability and uniformity has led even to a substantial role for methods usually associated with the hard sciences, as witnessed in the rise of randomized controlled trials in development economics.

This solutions-oriented approach can now be said to occupy the mainstream of development studies, as several studies have found. The abovementioned analysis by Madrueño and Tezanos Vázguez (2018) concludes that development studies is "cross-disciplinary" rather than interdisciplinary and, moreover, dominated by development economics. Similarly, Mitra et al. (2020) found a low level of citations between articles from different disciplines concerned with development; the majority that did exist took place between articles in development economics, development studies, and economics. The disciplines that the authors considered in this study were "generalist" development studies, sociology, anthropology, economics, development economics, geography, and political science. History was not included.

On the other side of the divide stand more critical analyses that, as EADI puts it, "question the very meaning of development and the politics underlying the development enterprise" (Mönks et al. 2019, 227). Here, anthropology and critical human geography predominate, although heterodox economics also contributes. Such analyses make frequent references to certain historical phenomena—particularly colonialism and Cold War geopolitical rivalry—but primarily in order to deconstruct development, which they view primarily as a discourse (for an early and seminal example, see Escobar, 1984). As such, they remain as wary of meso-level analysis as their economist counterparts, with an added distrust of materialist explanations of historical socio-economic change. Ethnography and discourse analysis are popular methods of data collection, and self-described historians seem to be thinly represented.

I am using a broad brush here, of course, and there are notable individual exceptions to these general trends. But when analyzing the (inter) disciplinary underpinnings of the field, history *as a discipline* has been surprisingly marginal. None of the working groups of the DSA or EADI (each has 17) focus explicitly on history. Although both have featured groups that focus on post-colonialism and decolonizing development studies, there exist well-documented tensions between postcolonial social theory and development studies (Power, 2003) that have not much been obviated since Christine Sylvester referred to them as "two giant islands of analysis and enterprise [that] stake out a large part of the world and operate within it or with respect to it as if the other had a bad smell" (1999, pp. 703–704). Not least, there are limitations to refracting the history of development interventions entirely through the lens of colonialism and North-South relations. As the field of global history has shown, decentering the North in order to also examine historical South-South and intra-South dynamics can prove extremely fruitful (see, for example, Hatzky, 2015). Previous North-North interventions may even be understudied *qua* "big D development": the North today contains endless internal pockets of assimilated or partly assimilated populations once deemed undeveloped "others" by modernizing elites.

The relative neglect of historical perspectives in particularly the policy-oriented realm of development studies has at times earned it the reputation of being faddish and prone to hype (Hobbes, 2014). As an anonymous practitioner warned their aspiring successors in *The Guardian*, "the ecosystem of aid and development entities is in a constant state of evolution … Today's brilliant innovation will be tomorrow's old hat. And the practice that you so passionately evangelize this week could well be proven harmful the next" ("J", 2016). Every field has paradigms that rise and fade from popularity. But because development policymakers and practitioners must "sell" interventions to distant parties as discrete packages that center around ever-evolving "theories of change" (modernization, import-substitution, the developmental state, structural adjustment, the green economy, and so forth), it is perhaps particularly noticeable here.

Efforts to bring kaizen to Ethiopia are prone to exactly such accusations of faddishness. And, indeed, within the few years that I have been studying the topic, political upheavals have transformed Ethiopia. An ideological re-orientation, two changes in Prime Minister, a civil war, and a pandemic (not to mention several more gradual shifts on the Japanese side) have all combined to put the future of the entire project in peril. At the time of writing, few of the goals set out by the EKI seem close to being met on time, or perhaps at all.

I would like to argue here, however, that the travails of this intervention do not mean it is not worth examining more deeply. In addition, for me it turned out to be worth studying not just from more traditionally development studies-oriented disciplines such as development economics, anthropology, or even political science but from the vantage point of history.

## Where Other Disciplines Were Helpful and Where They Fell Short

In making the methodological and theoretical choices that would inform my study, I was confronted with possibilities from a range of disciplines. I drew on these to varying degrees but ultimately found the most useful contributions in the works of historians—an unexpected development.

From a *development economics* perspective, Ethiopian kaizen should be judged primarily on its ability to increase the productivity of target firms across entire sectors. Depending on how successful this particular intervention is deemed to be, the prescription will be either to tweak or to abandon it in other African countries. This research aim is not unimportant, but it is difficult to do in the Ethiopian context; only much more limited impact assessments have been undertaken (e.g., Getahun Tadesse, 2018). It also delivers a rather incomplete picture. Such assessments cannot tell us why kaizen is implemented, nor how it interacts with the existing worldviews of stakeholders and the institutions in which they are embedded.

*Political science* and *critical human geography*, with their interest in how ideas about development diffuse among or travel between polities, could be drafted in to answer some of these questions. Each has, in recent years, created sophisticated typologies and theories to explain (in the case of the former) and critique (in the case of the latter) these processes. Some of these theories proved valuable in my previous research into Ethiopian and Kenyan elites' engagement with East Asian "models" of development. These elites, I found, selectively emulated those East Asian countries whose historical trajectories and cultural contexts they perceived as similar to their own (Fourie, 2014, 2015). Thus, concepts such as lesson-drawing, cross-societal emulation, policy assemblages, policy mobilities, and policy transfer provide a vocabulary through which we can talk about traveling ideas and demonstrate their prevalence. But in this case, again, I encountered limits to the usefulness of such theoretical frameworks. I was not aiming to contribute to covering laws explaining the conditions around which certain types of transfer take place. On the other hand, critical human geography, with its more constructivist and contingent approach, seemed to demand first and foremost an engagement with scale and spatiality, as well as an a priori opposition to the "political-economic construction of neoliberal globalisation" (McCann & Ward, 2013, p. 8). Without taking a position on the feasibility of such covering laws or the flaws of neoliberalism, what I aimed primarily to understand was the contribution of this particular intervention (namely kaizen in Ethiopia) to the evolution of development theory and its prescriptions for socio-economic progress.

The *anthropological* perspective can very fruitfully contribute to answering these questions, and, indeed, it informed many of my methodological choices. In order to understand how and why kaizen was being implemented in Ethiopia, I drew most of my data from qualitative interviewing and ethnography. I conducted participant observation in Ethiopian factories, Japanese factories, study exchanges between Japanese and Ethiopian experts, and conferences dedicated to the dissemination of kaizen throughout Africa. I also conducted critical discourse analysis of Ethiopian and Japanese documents and spoken discourses.

Anthropology's utility went beyond the methodological. My previous research on Ethiopia had acquainted me with the influential concept of high modernism, coined by the anthropologist James Scott. High modernism, as defined by Scott, is

> a strong, one might even say muscle-bound, version of the self-confidence about scientific and technical progress, the expansion of production, the growing satisfaction of human needs, the mastery of nature (including human nature), and, above all, the rational design of social order commensurate with the scientific understanding of natural laws. (2008, p. 4)

It is, in Scott's and my usage, both a mode of governmentality and an ideology of intervention in the affairs of a society or target population. It is uncompromisingly top-down, almost by definition detrimental to the populations it claims to want to "help" (the subtitle of Scott's seminal book on high modernism is *How Certain Schemes to Improve the Human Condition Have Failed*) and most often carried out by the state. It completely ignores *metis*, "the knowledge that can come only from practical experience" (p. 6).

Although high modernism originated in Europe and went on to appear on almost every continent in the twentieth century, Ethiopia under the authoritarian Derg regime (1974–1991) serves as one of Scott's case studies. In my previous research on Ethiopia, I, too, had discovered strong high modernist undercurrents that spanned the pre-Derg and post-Derg eras. Drawing on the works of political scientists such as Christopher Clapham (2006) and historian Bahru Zewde (2002), I soon realized the important role that top-down, ambitious projects of statebuilding played for modernizing elites both envious and (rightly) suspicious of Western imperialism.

## Enter "Low Modernism" and Post-war Japan

Research articles in the social sciences typically present empirical enquiry as a logical, step-by-step process in which the research question, theoretical framework, and methodological choices proceeded smoothly from one another. In my experience, the reality is frequently far messier— although the reader would be hard-pressed to discover this from the finished journal article, book, or dissertation. In my case, the serendipitous connection between industrial productivity, high modernism, foreign development aid, and policy emulation came about through a combination of wide reading and intuition. Although I began with the suspicion that kaizen transfer represented not only the introduction of a set of workplace practices to a new setting but also a deeper change in the assumptions and prescriptions of global development agendas, what exactly this deeper change entailed was not initially clear to me. Reading widely and historically provided me with theoretical concept I had been missing, namely *low modernism.*

My discovery of high modernism had been so exciting to me because it chimed with a phenomenon that I and many other development scholars have long observed: that the spread of development interventions has long been intimately tied to the spread of modernity. Very often, this process has happened in an extremely top-down fashion. Witness Europe's incursions into its various colonies, the United States' promotion of modernization theory during the Cold War, and China's assimilation of Tibet. Contemporary discussions around the introduction of ambitious infrastructural projects, technologically intensive farming methods, export-processing zones, programs to assimilate pastoral populations, and sophisticated surveillance technologies in the Global South echo these concerns today, particularly when they are enacted by the state and supported by international finance. At the same time, there has been a surge of interest in "bottom-up" interventions that are meant barely to resemble interventions at all, supposedly providing local stakeholders with the space and resources to exercise their own agency and realize their own self-defined goals.

The more I learned about kaizen and its use as a modality of foreign aid, the more it occurred to me that this resembled something between these two approaches. As I have summarized it:

> In kaizen, a range of practical and visual features exist to encourage workers at all levels—but particularly on the shop floor—to continuously reflect on and improve their daily operations. Groups of frontline workers also meet regularly in order to suggest improvements to management, and this is

claimed to increase not only their efficiency but also their job satisfaction and sense of ownership. (Fourie, 2020, p. 1)

My research on the way the methodology is taught by Japanese and Ethiopian consultants in Ethiopia demonstrated a Janus-faced quality. Without replicating here in full the argument put forth in my original article, I found kaizen and its proponents to be deeply concerned with changing the mindsets of frontline workers toward a newfound appreciation for scientific rationality, productivity, and efficiency. But this, they emphasized, was only half of the equation. Simultaneously, the existing expertise of these workers must be elicited and taken seriously; managers must learn to respect workers' intimate, first-hand knowledge of the factory floor and reorganize power structures accordingly. It was thus only through this combination of modernity's disciplinary and participatory logics that industrialization could take off in Ethiopia.

To many workers in Western post-industrial settings, these workplace prescriptions may seem unremarkable at best or regressive at worst. Within the strict hierarchies of most Ethiopian factories, however, they hold a more revolutionary potential. And within the landscape of current international development cooperation, they are equally notable. Few donors have recently been comfortable articulating a desire both for top-down and bottom-up development, with the dividing line usually running between emerging donors (such as China and South Korea) and traditional Western donors whose own industrial development is by now distant memory.

It is within this context, then, that the final piece of my theoretical puzzle fell into place. The question we as supervisors in several branches of the social sciences often pose to students who have found an interesting real-world puzzle or phenomenon is: "what is this a case of?" My own answer to this question was that the programme of kaizen promotion enacted by the Japan International Cooperation Agency (JICA) in Ethiopia was not a case of high modernism but rather a case of *low modernism.*

Low modernism is not a concept used in any of the disciplines discussed in the previous section. Instead, it has been coined and developed by a handful of historians to describe certain nineteenth- and early

twentieth-century programs of transnational and national modernization. Again, quoting from my original article:

> Gilbert (2003) has coined the term "low modernism" to describe the efforts
> of agrarian economists to increase agricultural outputs in the United States
> in the 1920s through massive engagement of farmers and rural citizens …
> McVety (2008) has highlighted how the low modernist visions of American
> agricultural extension programs in Ethiopia in the 1950s clashed with the
> high modernist vision of Haile Selassie. Fischer-Tiné (2018) has demon-
> strated how American missionary-run rural reconstruction projects in
> interwar colonial South Asia sought to combine "self-help with intimate,
> expert counsel," thereby creating complex, multilevel epistemic communi-
> ties incorporating villagers and other subaltern populations. (Fourie,
> 2020, pp. 3–4)

By introducing low modernism into discussions of twenty-first-century
foreign aid, I was able to contribute to a genealogy of development interventions. Japan—neither quite an emerging nor quite a traditional
donor—emerged as the inheritor of those Western modernizers who
once themselves had to juggle the competing demands of democratization and industrialization abroad and at home. In addition to historiciz-
ing international development theory, I was also able to theorize history.
High modernism and low modernism differed in the answers they gave
to two key questions (Fig. 1), I argued: (1) *Who has the necessary expertise
and legitimate mandate to enact the modernization of a society?* And (2) *how
fast and how far should the process of modernization proceed?* Low modernism advocates for the inclusion of non-elites as regards the first question
and urges for some degree of caution as regards the second. This was a
more formal operationalization than had yet been given to the concept of
low modernism and has opened it up to further historical applications.
The interdisciplinary cross-fertilization thus worked in both directions.

Before closing this section, it is important to note one further important role that historical studies played in my analysis. The deeper I moved
into exploring kaizen and its precepts, the more I realized I could not do
this justice without also exploring the history of kaizen's emergence and
development in post-war Japan. Like all social concepts, kaizen is, after

**Fig. 1** Low modernism shares high modernism's faith in science, planning, and rationality but seeks to temper these with a recognition of the value of local knowledge (*metis*) and more iterative, gradual processes. © Fourie

all, a construct of human actors acting in path-dependent but agential ways. Here history's contribution was not methodological or even theoretical but rather empirical: it was only by reading historians' analyses of post-war Japan that I was able to understand how kaizen had been used both to control and to incentivize workers during this most tumultuous period of the country's history. It may sound strange to frame my use of this collection of secondary sources (for the most important, see Gordon, 1998) as a contribution from history. After all, scholars from most fields of study are encouraged to include a historical background section to their case studies. But it is my experience that studies of development cooperation do not commonly delve into processes of historical socioeconomic development in donor countries nor explore how exported "development lessons" first manifested in the countries that claim to have invented them. This is even more true of donors like Japan, which sits uneasily within the postcolonial dichotomies of critical development scholars and the post-war poverty of which is often seen as too distant for mainstream development economists.

The result of my drawing on these histories of Japan's own historical industrial relations was not only a more fine-grained understanding of kaizen's origin but a key conclusion: kaizen was the result of a grudging and difficult compromise between Japanese trade unions and American-backed industrialists. This suggests that low modernism can grow from a range of different national soils. Foreign intervention played a role but not the defining one. More importantly, kaizen is not a set of natural laws or an abstract theory of change designed by social scientists or management experts. Rather, it is a *political settlement*—an insight that Japanese donors, Ethiopian recipients, and would-be low modernizers elsewhere would do well to keep in mind. The fact that kaizen emerged in a foreign setting does not mean it cannot work in Ethiopia, but embedding it in broader discussions of domestic labor relations and power hierarchies will give it a better chance of succeeding.

## Conclusion: Looking Back to Look Forward

This chapter has told the story of how I came to study the transfer of kaizen to Ethiopia, and how, in the process, I came to realize the indispensability of certain historical debates and concepts. In part, this was a function of casting the net of my literature search further into the past than I had first anticipated, to 1940s Japan. It was, however, also due to the conceptual innovation of historians. This meant that I not only had to travel to post-war Japan but also had to the American Midwest in the 1930s, southern India in the 1920s, and Ethiopia's own early entanglements with foreign aid in the 1950s. One concept—low modernism—united these seemingly unrelated settings and, I realized, also existed unidentified in a modern setting. Kaizen is a slippery concept, and it means many things to many different people. It is only by viewing it in relation to other forms of foreign aid, other ways of stimulating industrialization, and other ways of positioning workers vis-à-vis management that its true significance as low modernist intervention becomes clear. This would not have been possible without the contribution of history.

In telling my story, I have also sought to contribute to a larger conversation around the identity and purpose of development studies as an interdisciplinary field. Here, I want to end with a modest plea toward my fellow development studies scholars. Stories matter, and history is one of the best collections of stories that we have at our disposal. Narratives are continuously being constructed and reconstructed, but we have histories for this too: histories of ideas. If development studies takes these histories seriously, we will better understand the contingent nature of our own ideologies of intervention and progress. This is the case whether we oppose or support contemporary development cooperation. Just as *metis*—the knowledge we can only gain through direct, intimate, and practical experience—helps to temper modernism and give it a crucial human dimension, getting our hands dirty in digging up dusty historical data and accounts can help to ground development studies. Here history's "middle of the road" approach—somewhere between the causal certainties of development economics and the critical deconstruction of anthropology and critical geography—may actually not be a sign of weakness, but rather strengthen our field. This also has implications for teaching as well: perspectives like global history and the history of science can play an important role in broadening the traditional foci of development studies and other social science degrees.[1]

I close, therefore, by explaining the metaphor of the frog from which this chapter draws its title. An idiom, sometimes attributed to Chinese and sometimes to Japanese historical sources, states that "the frog in the well knows little of the sea." The vast ocean of historical experience has certainly enriched my understanding of contemporary development policy in ways that I could not have anticipated from the confines of the well.

---

[1] For example, in the MA Globalisation and Development Studies and the BA Global Studies programs at the University of Maastricht, historians teach courses on the history of globalization and the history of sustainability, respectively.

# References

Clapham, C. (2006). Ethiopian development: The politics of emulation. *Commonwealth & Comparative Politics, 44*(1), 137–150. https://doi.org/10.1080/14662040600624536

Currie-Alder, B. (2016). The state of development studies: Origins, evolution and prospects. *Canadian Journal of Development Studies/Revue canadienne d'études du développement, 37*(1), 5–26. https://doi.org/10.1080/02255189.2016.1135788

Dados, N., & Connell, R. (2012). The global south. *Contexts, 11*(1), 12–13. https://doi.org/10.1177/1536504212436479

Development Studies Association (DSA). (n.d.). *What is development studies?* https://www.devstud.org.uk/about/what-is-development-studies/

Escobar, A. (1984). Discourse and power in development: Michel Foucault and the relevance of his work to the Third World. *Alternatives, 10*(3), 377–400. https://doi.org/10.1177/030437548401000304

Fischer-Tiné, H. (2018). The YMCA and low modernist rural development in South Asia, c.1922–1957. *Past & Present, 240*(1), 193–234. https://doi.org/10.1093/pastj/gty006

Fourie, E. (2014). Model students: Policy emulation, modernization, and Kenya's vision 2030. *African Affairs, 113*(453), 540–562. https://doi.org/10.1093/afraf/adu058

Fourie, E. (2015). China's example for Meles' Ethiopia: When development 'models' land. *The Journal of Modern African Studies, 53*(3), 289–316. https://doi.org/10.1017/S0022278X15000397

Fourie, E. (2020). Humanizing industrialization?: Japanese productivity methods, Ethiopian factories, and low modernism in foreign aid. *Global Perspectives, 1*(1), 17426. https://doi.org/10.1525/gp.2020.17426

Getahun, T. M. (2018). Kaizen as policy instrument: The case of Ethiopia. In K. Jin & K. Otsuka (Eds.), *Applying the kaizen in Africa* (pp. 151–198). Palgrave Macmillan.

Gilbert, J. (2003). Low modernism and the agrarian new deal: A different kind of state. In J. H. Adams (Ed.), *Fighting for the farm: Rural America transformed*. University of Pennsylvania Press.

Gordon, A. (1998). *The wages of affluence: Labor and management in postwar Japan*. Harvard University Press.

Gore, C. (2000). The rise and fall of the Washington consensus as a paradigm for developing countries. *World Development, 28*(5), 789–804. https://doi.org/10.1016/S0305-750X(99)00160-6

Hart, G. (2001). Development critiques in the 1990s: *Culs de sac* and promising paths. *Progress in Human Geography, 25*(4), 649–658.

Hatzky, C. (2015). *Cubans in Angola: South-South cooperation and transfer of knowledge, 1976–1991*. University of Wisconsin Press.

Hobbes, M. (2014, November 18). Stop trying to save the world. *The New Republic*. https://newrepublic.com/article/120178/problem-international-development-and-plan-fix-it

"J". (2016, September 20). Want to change the aid industry? Here's how to do it. *The Guardian*. https://www.theguardian.com/global-development-profes-sionals-network/2016/sep/20/how-change-aid-industry-humanitarianism

Madrueño, R., & Tezanos Vázguez, S. (2018). The contemporary development discourse: Analysing the influence of development studies' journals. *World Development, 109*, 334–345. https://doi.org/10.1016/j.worlddev.2018.05.005

McCann, E., & Ward, K. (2013). A multi-disciplinary approach to policy transfer research: Geographies, assemblages, mobilities and mutations. *Policy studies, 34*(1), 2–18.

McVety, A. K. (2008). Pursuing progress: Point four in Ethiopia. *Diplomatic History, 32*(3), 371–403. https://doi.org/10.1111/j.1467-7709.2008.00698.x

Mitra, S., Palmer, M., & Vuong, V. (2020). Development and interdisciplinarity: A citation analysis. *World Development, 135*(105076), 1–17. https://doi.org/10.1016/j.worlddev.2020.105076

Mönks, J., Carbonnier, G., Mellet, A., & de Haan, L. (2019). Novel perceptions on development studies: International review and consultations toward a renewed vision. In I. Baud, E. Basile, T. Kontinen, & S. von Itter (Eds.), *Building development studies for the new millennium* (pp. 217–241). Palgrave Macmillan.

Power, M. (2003). *Rethinking development geographies*. Routledge.

Scott, J. C. (2008). Seeing like a state: How certain schemes to improve the human condition have failed. Yale University Press.

Sylvester, C. (1999). Development studies and postcolonial studies: Disparate tales of the 'third world'. *Third World Quarterly, 20*(4), 703–721. https://doi.org/10.1080/01436599913514

Zewde, B. (2002). *A history of modern Ethiopia, 1855–1991*. Ohio University Press.

# Part II

## Refolding Methods: How Twists Require Tweaks

# Examining Personal and Cultural Narratives of Aging: Literary Gerontology Revisited

Aagje Swinnen

## Introduction

The academic discipline of literary studies *matters* because it elucidates how literature intervenes in and shapes the social world. The type of cultural work performed by literary texts helps us to imagine a future that is more inclusive of women, LGBTQ+, older people, people of color, people who live with a disability, and nonhuman animals. Some have even argued that literature makes us better citizens. Martha Nussbaum (1990), for instance, believes that great literary fiction can generate moral insight. Such convictions about the role and benefits of literature are part of the

I would like to thank the reading and writing club participants, Annette de Bruijn, as well as colleagues who have commented on an earlier draft of this text, especially Anita Wohlmann. I also thank Michele Faguet for proofreading the text.

A. Swinnen (✉)
Department of Literature and Art, Faculty of Arts & Social Sciences, Maastricht University, Maastricht, The Netherlands
e-mail: a.swinnen@maastrichtuniversity.nl

© The Author(s) 2023
K. Bijsterveld, A. Swinnen (eds.), *Interdisciplinarity in the Scholarly Life Cycle*,
https://doi.org/10.1007/978-3-031-11108-2_8

credo that informed my own academic formation. In this chapter, however, I will recount how I grew uncomfortable with this and what steps I took to reinvent my scholarly practice. I will detail the methodological innovation with which I experimented, the results, and how this eventually helped me to reevaluate my point of departure as a literary scholar specialized in representations of aging and later life.

## Literary Gerontology and Oppositional Reading

As Sarah Falcus has written, "in literary studies, ageing has been the unacknowledged shadow that intersects with more prominent approaches such as gender or postcolonialism" (2015, p. 53). This has been changing since the 1980s, both due to the influence of feminist literary criticism, evident in my own, early work (Swinnen, 2006), and the rise of cultural studies. They paved the way for literary scholarship focused on issues of representation related to aging and later life, a field referred to as literary gerontology. Literary scholars who practice literary gerontology engage in a type of oppositional reading that reveals the often hidden age ideologies of a text while simultaneously addressing points of exit from these ideologies. As such, they examine the power of the text "to demystify, destabilize, denaturalize" as well as "to recontextualize, reconfigure, or recharge perception" (Felski, 2015, p. 17). The work of literary gerontologists is ethical and political. It aims to clarify cultural meanings of aging, negotiating and subverting them in a world characterized by structural and everyday ageism. There is a clear preference within literary gerontology for dialogic highbrow literature that includes different points of view on the aging experience, which often intersects with the experience of other crucial differences. The underlying idea is that being confronted with these points of view has an educational effect on the reader and thus society at-large.

At a certain point, however, I felt a growing discomfort with the claims of these reading practices, and this did not change even after I broadened my scope from representations of aging and later life in literary texts to photography, film, television, and performance, in the interdisciplinary setting of Maastricht University. I kept wondering whether the "de-" and "re-"construction of meaning in a variety of cultural texts could really advance social justice for older people who are, by and large, positioned as *the* problem of population aging. Despite engaging in outreach work through conventional formats such as public presentations or publications in periodicals, I became dissatisfied with the lack of scholarly connection between the images of fictional aging and the experiences of aging people around me. Analyses of representations of older characters in texts seemed far removed from the challenges that older people are faced with, especially the persistent pressure to age successfully by staying young forever—an expectation that many have internalized.

How could I not just write *about* older people but also collaborate and co-create knowledge *with* them? How could I connect academic discussions of artistic value and meaning with ordinary life and lay readings? I felt that cultural gerontology had something to offer in this regard.

## Narrative as a Concept that Connects Literary and Cultural Gerontology

While literary gerontology refers to literary studies that implement age as a critical perspective, cultural gerontology encompasses social science approaches to the study of old age influenced by the cultural turn. Cultural gerontology builds on the notion that medical and chronological definitions of age are neither static nor fixed (Twigg & Martin, 2015, p. 2). For example, neoliberalism and consumerism have profoundly altered cultural and personal notions of aging in late modernity. The positioning of older people as a demographic that is a cause of alarm—a threat to the welfare state—originated in the nineteenth century (Katz, 1992), while the rise of third-age lifestyles to maintain health and productivity as long as possible is a product of the 1980s (Gilleard & Higgs, 2013). In order to move away from old age as a problematic category, cultural gerontology prioritizes the creation of "a fuller and richer account of later years … one that places the subjectivity of older people, the width and depth of their lives, at the forefront of analysis" (Twigg & Martin, 2015, p. 2). This phenomenological interest in the exploration of people's subjectivities in combination with the metaphor of life as a story is among the reasons that cultural gerontology offers opportunities to build bridges with the humanities and literary studies of aging and later life. "Narrative" seems to be the magical concept that facilitates such exchanges (Falcus, 2015; Swinnen & Port, 2012). There is a broad consensus about the storied nature of human experience in both fields. People make sense of the world through narratives and use narrativity to constitute their identities, although we should be careful not to overemphasize narrative coherence for understandings of personhood since this can be detrimental to people who live with dementia, for example. Some would argue that cultural gerontology is influenced not only by the cultural turn but also by the narrative turn in the social sciences, which resulted in the emergence of narrative gerontology. I feel that it would be too far-reaching to delve into the distinction between cultural, narrative, and literary gerontology (and critical gerontology by extension) in such a short chapter. I would recommend Kate de Medeiros (2014) and Hannah Zeilig (2011) for more nuanced accounts; both scholars work at the intersection of the humanities and social sciences.

For a cultural gerontologist, narratives offer a specific type of knowledge that can be elicited in numerous ways, for instance, by means of qualitative interviewing, oral history, or ethnographic approaches that include participant observation and photovoice. Experiences and opinions that older people are able to voice through narrative do not exist in a vacuum, however. They are connected both to the larger master narratives (values and ideologies) about aging that circulate in our society and to the communities of people that we engage with and depend on to generate meaning. Consequently, what people articulate during an exchange with an interviewer or in the presence of a field worker is not necessarily an unmediated reflection of their private opinion. For this reason, it is also important to listen to and detect what cannot be spoken and to understand that during the processes of transcription and note-taking, narratives become disembodied and distanced from the moment in which they emerged (de Medeiros, 2014, p. 34). Another factor is that internalized ageism may make interviewees reluctant to talk about old age or they may verbally distance themselves from the topic by rehearsing normative age scripts that value individual agency and control, independence, productivity, agelessness, and permanent personhood. The cultural gerontologist, in contrast, hopes to move beyond these scripts through their research design. It is in this realm that the interpretative skills and inventiveness required of the cultural gerontologist are comparable to those of the literary scholar.

For a literary scholar, a narrative is not just any story but a narrator's particular representation of a series of events which formal features or aesthetics are crucial to its meaning. Literary scholars are trained to situate literary works in the context of their production and to illuminate the ambiguities of fictional narratives, including life histories such as (auto) biographies and memoirs. Rita Felski (2015) has raised the question of whether some types of literary critique—and cultural critique in general—go too far by looking for hidden meanings in a text, which is, in a sense, the opposite extreme of taking texts at face value (cf. de Medeiros). Felski argues that the almost antagonistic attitude toward the literary artifact goes hand in hand with a disregard for ordinary readers "who persist in using these texts in unseemly or inappropriate fashion—identifying with characters, becoming absorbed in narratives, being struck by moments of recognition" (2015, p. 29). In other words, the antagonistic attitude would imply that both the ordinary reader, who presumably takes the text at face value, and the literary text itself are "oblivious to its … latent contradictions" (p. 66).

If connecting ordinary readers, personal life experiences, and embodied responses with literary texts is almost taboo in literary studies, how can we combine the study of narratives in relation to the lived experiences of older people in cultural gerontology with the oppositional readings of literary gerontologists? How can we bring the readings of so-called lay people and professionals together? I found great inspiration and some answers in the Fiction and Cultural Mediation of Ageing Project (FCMAP) at Brunel University.

# The Methodology Behind FCMAP

A team of scholars affiliated with the Brunel Centre for Contemporary Writing conducted the FCMAP from May 2009 to February 2012. FCMAP examined "(1) the relationship between cultural representations of, and social attitudes to, ageing and (2) the potential of critical reflection and elective reading by older subjects for engendering new ways of thinking about ageing" (Hubble & Tew, 2013, Chapter 1, para. 1). It did so by means of what Nick Hubble and Philip Tew called "an innovative methodological bricolage" (Chapter 1, para. 1) that involved collaboration with the social research organization Mass Observation (MO), the Third Age Trust, and the think tank Demos. In my view, FCMAP's methodological approach brought literary and cultural gerontology together in two ways. On the one hand, it made use of close readings of literary texts combined with an analysis of literary authors' statements on the topic of old-age representations. These statements were elicited during a series of literary events staged specially for the project. On the other hand, FCMAP combined these "literary voices" with ways to collect, elicit, and analyze narratives of everyday life by collaborating with MO and the Third Age Trust.

Founded in 1937, MO is known for employing participatory research techniques to gauge public opinion in Postwar UK, for instance, through the combination of questionnaires and diaries that are kept over a longer period of time following clear instructions. FCMAP made use of the longitudinal qualitative data on aging already generated by Pat Thane in 1992 and 2000 via MO as well as 193 responses to newly developed instructions concerning representations of aging in political and media discourse. Additionally, in collaboration with the Third Age Trust, FCMAP set up eight groups with 80 volunteers in their early 60s–90s. Over the course of one year (2009–2010), the groups read nine novels published from 1944 to the present. These included, for example, Hanif Kureishi's *The Body* (2002), Angela Carter's *Wise Children* (1991), and Barbara Pym's *Quartet in Autumn* (1977). The participants kept diaries about the books and the group discussions; some of them attended literary events. Demos made the findings of the project available to policymakers in a 200-page report, *Coming of Age* (2011), which was also presented at several events to reach stakeholders. Many of them felt that the elicited narratives provided valuable information on age-related issues usually dealt with in a more top-down fashion (Hubble & Tew, 2013, Chapter 1, para. 9).

## Experimenting with a Reading and Writing Club in Maastricht

In preparation for my inaugural lecture (Swinnen, 2017) as Endowed Socrates Chair in Humanism and the Art of Living at the University of Humanistic Studies, I decided to take a leap and pursue a modestly scaled project modeled after FCMAP. I wanted to experiment with data collection by forming a reading club (Fig. 1), particularly since I had already

I started to look at myself through his eyes. How do others see me now? For instance, when I walk somewhere. Are you still capable of walking these stairs? We all age, of course. We can't do anything about it. But that you are confronted with it in such a nasty way and then have to resist it...

**Fig. 1** Collage based on an illustration by Janneke Swinkels that was inspired by a photograph of the fieldwork with the reading club in 2017 © Swinnen

been studying cultural representations of aging in a variety of texts. My aim was to examine (1) reading experiences of people over 60 in the context of their lived experiences and (2) processes of critical reflection on aging, self, and society through reading, writing, and exchange. I launched a call for participation at the end of a public lecture that I gave for Vlam, a Maastricht organization specialized in literary events. After I informed ten aspiring participants about the work the project would require, eight signed up: six women and two men with an average age of 71, the youngest 61 and the oldest 82. All were highly educated except for one person who described his professional life as "many jobs that required manual rather than intellectual labor." Two members were published authors of fiction, but rather than singling them out as FCMAP did, I had them work together with the other readers in a less hierarchical way.

For the data collection, for which I sought approval from the Ethical Review Committee Inner City of Maastricht University, I used two sets of questionnaires that participants filled out at the beginning and end of our collaboration. Examples of the questions posed included: What does aging mean to you? To what extent has participation in the book club changed your perception of later life? To what extent have you experienced sharing reading responses as enriching and why? Like the FCMAP volunteers, participants kept diaries throughout the reading process. The books that the reading club discussed included different types of prose around topics such as love in later life, elder suicide, late-life creativity, care and institutionalization, and dementia of a parent. We started with five literary works in Dutch, including Dimitri Verhulst's *Madame Verona Comes Down the Hill* (*Mevrouw Verona daalt de heuvel af*, 2006) and Erwin Mortier's *Stammered Songbook: A Mother's Book of Hours* (*Gestameld liedboek: Moedergetijden*, 2011), and ended with three books originally written in English, which some of the participants read in translation, for example, Elisabeth Strout's *Olive Kitteridge* (2008). As the collaboration progressed, I gave book club members more of a say in the selection of the works. I asked participants to include any responses in their reading diaries that seemed of relevance to them, including how the literature evoked certain thoughts, recollections, and emotions.

Thus far, my methodological approach was fairly similar to FCMAP's collaboration with volunteers through the Third Age Trust. However, I also made some changes, not only because I had a limited time span in which to run the project, but because I felt that I could further develop the FCMAP approach. Instead of the more unstructured book club meetings that Hubble and Tew (2013) preferred "[to minimize] the influence of researchers upon respondents" (Conclusion, para. 2), I chose a more structured approach modeled after a focus group interview. The underlying idea, nonetheless, remained that reading is a particular social event and that the exchange of reading experiences may deepen reflection on topics such as aging. I did not contribute as a reader to the focus group interview but chaired and moderated the discussion by asking open questions, for example, "What do we learn about the main character of the book?" and "Which excerpts would you single out as surprising or especially meaningful?" I made it clear to the participants that any response was welcome and that the aim of the conversation was not to arrive at a consensus. I refrained from providing my own interpretations of the novels even when members of the group explicitly asked for my expert opinion. The participants did not reflect on the group discussions in their reading diaries. Instead, all discussions were recorded and transcribed verbatim by my colleague Annette de Bruijn. This guaranteed a great amount of detail in the data on group discussions.

To further obtain information on how the participants experience aging and assess representations of aging, I added creative writing exercises to the questionnaires, reading diaries, and group discussions. My hypothesis was that creative writing exercises are another way to disclose hidden attitudes to aging as well as show participants' reactions to how aging is represented in the literary texts. The participants were free to write about anything they wanted, but I also gave some options. For example, in relation to the novella *Madame Verona Comes Down the Hill*, which ends with the main character committing suicide by staying out in the freezing cold all night, I suggested rewriting the ending. One reader chose to let the protagonist live (Nono);[1] another introduced the perspective of a daughter coming to terms with Verona's death (Cunera). In their questionnaires, participants noted that the creative writing exercises "appeal to your creativity and make you think about how and why the

---

[1] The names of participants have been changed to guarantee anonymity. All translations are mine.

author chose and elaborated a particular storyline" (Aspirant) and they "force [you] to order [your] thoughts about what was written" (Simone). We used Dropbox to share all the reading diaries and creative writing exercises. Some participants also used Dropbox to upload reviews of the literary works and interviews with the authors in order to contextualize the literature list. I analyzed the entire data set through several rounds of close readings, as literary scholars are trained to do.

## Some Findings About Experiences of and Attitudes Toward Aging

The primary aim of my experiment was to have a better understanding of how older people experience aging and what they think of aging. To do this, I engaged a new methodological approach consisting of an interactional setting that used literature as the departure point of conversation. Discussions about reading and writing assignments performed "a specific type of cultural work, for they enable participants to articulate or even discover who they are: their values, their aspirations, and their stance toward the dilemmas of their worlds" (Long, 2003, p. 145). My diverse data set definitely yielded a wealth of information in this regard that I will concisely summarize because it builds on the findings of Hubble and Tew. They end their book by optimistically speculating about

> the emergence of a new social narrative of ageing, which both allows for long active post-retirement years and a gradual acceptance of old age, seen not as decrepitude or [a] social problem, but as the attainment of a self-acceptance that transcends any purely medical concept of well-being. In other words our public and social concept of successful ageing has to be revised. (Hubble & Tew, 2013, Conclusion, para. 2)

My data shows more of the ambiguity of the Dutch participants toward aging and how they together arrive at a basic understanding of ageism.

In terms of self-perception, all participants felt younger than their chronological age. Some connected this feeling explicitly with the notion of the "ageless self," which is a rather problematic feature of the successful-aging paradigm because it suggests the presence of a stable self, unaffected by the experiences developed over time while trapped in an aging body (Gibbons, 2016). Cunera, for instance, writes in one of her questionnaires: "An older person is always still that much younger 20 or 30-year-old. When I meet older people, I am initially shocked by their grumpy, closed, worn-out appearance, but once I begin speaking with them, I always see the 30-year-old appear." To maintain youthfulness, the participants agree that certain measures must be taken. As Nono puts it, "I am hard on my peers. I feel that you have a duty to make an effort to stay healthy and strong and mentally fit" (questionnaire). This emphasis on the duty to take responsibility over one's own health and happiness is another troubling characteristic of the successful-aging paradigm. It denies the contingencies of life and the physical, relational, and existential vulnerabilities that people are confronted with throughout their life course. Nono continues with the warning that no one enjoys listening to bitter nostalgia or boring accounts of health issues. Evidently, this is what she associates with older people, a category to which she does not belong. Participants often mention dementia as the most horrifying and inhumane form of decline.

All these instances of stereotyping and distancing oneself from older people, especially those unable to live up to the ideal of successful aging as everlasting youthfulness, are indicative of what has been called internalized ageism. Although the participants showed sensitivity for other types of isms, such as sexism and racism, they were unfamiliar with the concept of ageism. This became most clear in their response to the international bestseller *The Secret Diary of Hendrik Groen, 83 ¼ Years Old* (*Pogingen iets van het leven te maken: Het geheime dagboek van Hendrik Groen, 83 ¼ jaar,* 2014), an example of what is called "geezer and grump lit" (Swinnen, 2019). In this fictitious diary, the protagonist, Hendrik Groen, narrates his experiences in a retirement home in Amsterdam-Noord. To fight the meaninglessness of his institutionalization, Groen founds the Old-But-Not-Dead Club together with six other residents. They organize activities and outings exclusive to the club's members, developing a sense of collective invincibility that is largely based on their dismissive attitude toward the home's other residents. Most of the participants in my project did not appreciate the way that Groen talks about the residents who are not part of his club. They were appalled by the way he positions them, even though some of them used similar language when discussing other older people, especially in their questionnaires.

Let me unravel a bit more what initially seems to be a contradiction between the group's response to Groen and their own attitudes toward older people. Vosje shared the impact of Groen's narrative on her own experience of aging:

> I started to look at myself through his eyes. How do others see me now? For instance, when I walk somewhere: are you still capable of walking down these stairs? And so on … [laughs uncomfortably]. You become more aware of your own aging, since we all age, of course. We can't do anything about it. But that you are confronted with it in such a nasty way and then have to resist it. (Group discussion transcript, May 29, 2017)

With these words, Vosje addressed the lookism that underlies ageism. Her fellow reading club members, however, did not identify with her words because they felt they had nothing in common with the older people in the book. Still, they agreed that there are limits as to what can be said about older people and the term "old" itself became a subject of scrutiny. Participants wondered if it should be replaced by a more inclusive term, just as the term "non-Western immigrant" (*niet-westerse allochtoon*) has recently been replaced by Moroccan-Dutch or Turkish-Dutch citizen (*Marokkaanse* or *Turkse Nederlander*), for example. Winterfall noted how we are often unaware of the prejudice lurking behind words:

> But such words—aren't they hidden deep within us? Not long ago, there was the item "hidden racism" in the media. People who, without even knowing it, buried all kinds of fascist ideas and words deep inside their heads that you, with the right incentives [laughs], bring out again …. (Group discussion transcript, May 29, 2017)

The group then began to discuss sexist speech in *The Secret Diary*, which they found difficult to swallow. Winterfall wrote in one of his questionnaires, "women are (still) judged more than men on how they look. And men (still) on their economic value." This idea returned during the group conversation. The dynamic among participants demonstrates how together and rather intuitively they arrived at a more nuanced and better understanding of the mechanisms of ageism and its intersection with sexism. However, it remained difficult for them to apply the terms "old" or "older" to themselves. This could suggest that reading fiction does not necessarily change internalized assumptions about aging or that older age is experienced in such diverse ways that readers do not automatically align themselves with older characters in fiction.

## Circling Back to Literary Studies

Although my initial focus was people's experiences of and attitudes toward aging, I quickly learned that the data I gathered and analyzed also revealed insights relevant to literary studies—reception studies to be precise—a field rarely connected to literary gerontology, to which I previously paid little attention.

Reception studies departs from the assumption that the meaning of a text does not reside in the literary work itself but emerges from the interaction between a text and a reader (Freund, 1987). It also claims that the background of each individual reader influences the production of meaning. Readers who share a similar background may be considered part of a specific "interpretative community" (Fish, 1980). What makes reception studies a forerunner to a postcritical approach in literary studies today (cf. Felski in this chapter) is its interest in ordinary readers, affective responses, and the potentially beneficial role of literature in people's lives. One particular tendency within reception studies is a feminist approach to women's reading practices. Janice E. Radway (1984), for instance, has shed light on women's consumption of romance novels, while Elizabeth Long (2003) has examined the reading practices of American women participating in book clubs. Radway was among the first scholars to unravel how book club members combine the reception of literary works with personal experience in a process of collective self-reflection that "enables self-discovery and collective affirmation" (Long, 2003, p. 146).

Although my reading and writing group, for people over 60 rather than just women, was established specifically for the research project, the data does reveal what values the participants ascribed to reading together. I could identify at least four. Firstly, they explained how a book club offers the opportunity to meet interesting and like-minded older people who prefer in-depth discussions to small talk (about ailments and grand-children). Secondly, the participants described the importance of being confronted with other people's responses and points of view with which to compare one's own experiences. Nono wrote, for instance:

> I enjoy my book clubs because we exchange our reading experiences. It is always very nice to hear other insights and to share your own with others. And, it is also wonderful to tell and hear about experiences that we recog-nize from our own lives. Such discussions can become heated, because they occasionally reveal views, norms, and values. (Questionnaire)

Some acknowledged that a reader's background, such as their education or profession, influences the reading experience. Thirdly, the participants claimed that such comparisons and exchanges of responses result in new insights that can be inspirational and enriching. This is what Long calls "intersubjective creation/accomplishment" (2003, p. 145). Cunera even addressed how a shared reading experience results in a type of catharsis:

> Alone is but alone. Only through interaction with others you become milder, stronger, more helpful, and are helped in return ... Through a book's message, catharsis takes place in a very safe, pleasant way. You can compare it to a room where you collectively listen to music. I suspect that discussing a book creates a more peaceful bond than watching a football game. (Questionnaire)

Lastly, being part of a reading community provided participants a sense of respect and belonging. Several readers indicated the importance of feeling appreciated and intellectually valued—more difficult as one ages—through participation in a book club.

From here it follows that participants first and foremost shared a "cog-nitive motive" (Duyvendak, 2005) for joining the reading group. The fondness for collective knowledge production may not be so surprising given that my participants consciously decided to engage with a research project. This begs the question as to what extent the project itself intervened in their lives—a question I was frequently asked when presenting the project to aging studies scholars who expected some kind of therapeutic effect. Most people reported in their final questionnaires that participation in the project itself did not really (or significantly) change their perceptions and experiences of aging. But they experienced a focus on representations of aging in prose as an eye-opener. Except for one of the published writers already familiar with my work prior to her involvement in the study, not one member of the reading group had ever paid attention to how the lives of older characters are depicted in literature and how they might be affected by these portrayals.

Their collective reflection on a selection of literary representations of aging gave insight into the role of recognition in this shared reading practice. I am not referring to the sudden joy experienced when discovering certain similarities with a character. Recognition is here understood as "acknowledgment," described by Felski as "a claim for acceptance, dignity, and inclusion" (2008, p. 29). Systematically looking into the ways aging is portrayed prompted the readers to come up with very clear ethical guidelines for authors to commit to representations that are more just. They recommended refraining from generalizations, which result in dangerous stereotypes and caricatures. As Simone formulated it in a questionnaire: "For example, if it is always written that physicality/sensuality diminishes as you get older, and you yourself have a different idea about it, that can give someone the impression that they are 'not normal.'" Cunera wrote: "I see too many caricatures in literature. It already begins with drawings for toddlers: grandmas and grandpas are all crooked, with glasses and a stick. All slow and too sweet and too understanding" (questionnaire). Instead, authors should pay attention to the differences between older people and how they experience aging.

The focus on the cognitive motive does not mean there was no interest in aesthetics (cf. Duyvendak's "aesthetic motives"). The participants were avid and experienced readers; most were part of other book clubs, some of which had already existed for 20 years. In general, they agreed on what a good reading and a good book entailed. Their unwritten literary views—rather similar to a literary critic's assumptions—were used as a yardstick.

Many readers, for instance, shared an interest in the author of the book, but they clearly distinguished between the narrator and the author. Knowing a little more about the author can shed new light on the work, though most readers agreed that the book itself is central rather than the author's intention or worldview. Simone gave the example of Louis-Ferdinand Céline who "wrote beautiful works but had despicable ideas about Jews" (questionnaire). Furthermore, project participants agreed that a literary work should not be confused with reality. At best, it offers a possible world in words that "offers an amplification of all variations of beauty, horror, and possible changes" (Cunera, questionnaire). There was also a tacit understanding of what a well-written book looks like: it offers enough gaps for the reader to interpret it in their own way. Even when they disapproved of the content, participants agreed that well-written books are a pleasure to read.

There was one outsider to this reading community (*not* the male participant with a different background) for whom enchantment trumped knowledge production. Her more hedonistic reading experiences—especially of the middlebrow novel *The Secret Diary*—were sometimes harshly judged by other readers. Still, this participant was equally vocal on how diverse and inclusive representations of aging characters should be—she was just more forgiving of stereotypes and norm-affirming humor.

## Conclusion

Hubble and Tew's "methodological bricolage" developed in the framework of FCMAP inspired me to embark on research activities—especially the gathering of data—departing from what is common practice in literary gerontology. It enabled me to rethink the value of ordinary readers' responses to fiction and question the hierarchical division between lay and professional readers. It also allowed me to discover the potential of reception studies to develop a postcritical approach to literary representations of aging and later life. Most importantly, though, it helped me to include older peoples' perspectives and experiences in humanities approaches to the study of aging. As I am finishing this piece, I have embarked on a new project called "Shared Reading in Times of Lockdown" in collaboration with De Culturele Apotheek (NL) and Bond zonder naam (BE). I will study the dynamics of reading groups spanning different generations that respond to poetry and prose on the topic of isolation. At the same time, there are projects emerging in Spain (Maricel Oró-Piqueras and Emma Domínguez-Rué), Sweden (Linn Sandberg and Karin Lövgren), Denmark (Peter Simonson), and Germany (Anita Wohlmann) that work with reading groups for people over 60 to discuss topics like resilience, late-life masculinity, and retirement. The future will determine how they will contribute to the innovation of literary and cultural gerontology, aging studies, and literary studies.

# References

de Medeiros, K. (2014). *Narrative gerontology in research and practice*. Springer.

Duyvendak, L. (2005). Gelijkgestemde zielen: Waarom vrouwen in groepsverband lezen. *Jaarboek voor Nederlandse boekgeschiedenis, 12*, 177–190.

Falcus, S. (2015). Literature and ageing. In J. Twigg & W. Martin (Eds.), *Routledge handbook of cultural gerontology* (pp. 53–60). Routledge.

Felski, R. (2008). *Uses of literature*. Blackwell Publishing.

Felski, R. (2015). *The limits of critique*. University of Chicago Press.

Fish, S. (1980). *Is there a text in this class? The authority of interpretative communities*. Harvard University Press.

Freund, E. (1987). *The return of the reader*. Routledge.

Gibbons, H. (2016). Compulsory youthfulness: Intersections of ableism and ageism in 'successful aging' discourses. *Review of Disability Studies: An International Journal, 12*(2–3), 70–88.

Gilleard, C., & Higgs, P. (2013). *Ageing, corporeality, and embodiment*. Anthem Press.

Hubble, N., & Tew, P. (2013). *Ageing, narrative and identity: New qualitative social research (Kindle version)*. Palgrave Macmillan.

Katz, S. (1992). Alarmist demography: Power, knowledge, and the elderly population. *Journal of Aging Studies, 6*(3), 203–225.

Long, E. (2003). *Book clubs: Women and the uses of reading in everyday life*. University of Chicago Press.

Nussbaum, M. (1990). *Love's knowledge: Essays on philosophy and literature*. Oxford University Press.

Radway, J. (1984). *Reading the romance.* The University of North Carolina Press.

Swinnen, A. (2006). *Het slot ontvlucht: De vrouwelijke Bildungsroman in de Nederlandse literatuur.* Amsterdam University Press.

Swinnen, A. (2017). *Goed ouder worden en creatief lezen: Het kunstje van Hendrik Groen.* Universiteit voor Humanistiek.

Swinnen, A. (2019). Reading ageism in "geezer and grump lit": Responses to *The Secret Diary of Hendrik Groen, 83, ¼. Journal of Aging Studies, 50,* 100794. https://doi.org/10.1016/j.jaging.2019.100794. Published online before print Jul 5, 2019.

Swinnen, A., & Port, C. (2012). Aging, narrative, and performance: Essays from the humanities. *International Journal of Ageing and Later Life, 7*(2), 9–15. https://doi.org/10.3384/ijal.1652-8670.1272a1

Twigg, J., & Martin, W. (2015). The field of cultural gerontology. In J. Twigg & W. Martin (Eds.), *Routledge handbook of cultural gerontology* (pp. 1–15). Routledge.

Zeilig, H. (2011). The critical use of narrative and literature in gerontology. *The International Journal of Ageing and Later Life, 6*(2), 7–37. https://doi.org/10.3384/ijal.1652-8670.11627

# Museology and Its Others: Analyzing Exhibition Storytelling Through Narratology, Space Analysis, Discourse Analysis, and Ethnographic Research

Emilie Sitzia

## Introduction

Two visitors push the heavy doors and enter the dark space of the exhibition. They stop, they hesitate, they look for an introduction panel but miss it as they are distracted by the mosaic of screens blaring at the entrance. On each screen a specialist is talking, but the sound is on for only one of them. It involves a commentary on Fernand Braudel's work by a distinguished urbanist. The visitors stand there a few seconds, look at each other quizzically and start moving again. There is an opening on each side of the screen, each one giving access to a different part of the exhibition. One visitor goes left, the other goes right. But then they stop, turn around, go back to each other, and try to determine which of the two paths is the correct path. Next, they notice a discrete map of the exhibition on a stand. After carefully

The research was undertaken during an Iméra/Mucem fellowship from February to December 2020. I want to thank both institutions for their support during this research period.

E. Sitzia (✉)
Department of History, Faculty of Arts & Social Sciences,
Maastricht University, Maastricht, The Netherlands
e-mail: emilie.sitzia@maastrichtuniversity.nl

K. Bijsterveld, A. Swinnen (eds.), *Interdisciplinarity in the Scholarly Life Cycle*,
https://doi.org/10.1007/978-3-031-11108-2_9

looking at it, they look back at each other, shrug their shoulders, and take the opening on the left. As it turns out, they entered the historical section, by chance rather than by choice. (Vignette drawing on fieldnotes Sitzia)

Museums are key players in constructing meaning, asserting individual and collective identities, and institutionalizing heritage. They also act as catalyzers in civil society and contribute to envisioning possible futures. As such, the narratives they put forward have a significant impact on how a particular society presents itself, perceives itself, and projects itself into the future. If museums aim to be inclusive and to act as agonistic spaces with layered multivocal and complex stories, sometimes they fail to communicate their narratives to their visitors.

Traditionally, to examine such communication of narratives, practitioners of museum studies have relied solely on visitor research, which is most often based on closed-question surveys and tends to give a very superficial, and sometimes biased, impression of the reception of the narratives. I propose here a mixed methods approach not only to analyze the nature of storytelling within the museum but also to assess whether those narratives translate into meaningful visitor experiences.

In recent decades, various studies and emerging practices have challenged the traditional, unidirectional educational and social role of the museum (Vergo, 1989; Sandell, 1998; Davallon, 1999; Mairesse & Desvallees, 2007; Dewey, 1916/2008; Marstine, 2006; Simon, 2010; McSweeney & Kavanagh, 2016; Antos et al., 2017; Janes & Sandell, 2019; Chynoweth et al., 2020). Using strategies popularized by "new museology" and "participative practices" (Vergo, 1989; Marstine, 2006; Simon, 2010), some museums have explored new pedagogical frameworks, alternative modes of building and exhibiting narratives, as well as audience-activating tools. These frameworks are meant to allow museums to engage publics of all ages, to be socially relevant to and inclusive of visitors from diverse social and cultural backgrounds, and to be representative of the multiple community voices in contemporary society.

Furthermore, in expanding the possible meanings of learning and knowledge (Sitzia, 2017, 2018), museums have become multimodal spaces of communication. In order to engage a variety of audiences and stimulate a wide range of knowledge production and skills, museums have developed a broad gamut of communication strategies. From traditional wall-texts and labels to video and audio-installations, interactive maps, smell vials, touch boxes, and dress-up chests, museums have become places of exploration, communicating elaborate and layered narratives in multisensory ways.

This implies that museums now have to take complicated decisions regarding the stories they choose to tell and the ways in which they tell them. Thus, according to Borg and Mayo, museums can be "conceived of as sites of struggle, of cultural contestation and renewal" (2010, p. 37). Indeed, museums attempting to challenge and question the monolithic national narrative are gradually becoming "agonistic museums," a term coined by Chantal Mouffe (2016). Similarly, others have addressed how museums may turn into institutions that "trouble identity, decolonize, mock, revisualize, tell alternative stories, reorient authoritative practice, interrogate intolerance and privilege and stimulate critical literacies" (Clover, 2015, p. 301). Now that many museums are willing to critically engage with the public and actively commit themselves to particular social issues, the narratives they present have become both more sophisticated and more layered.

In turn, this situation requires from us that we adapt the ways we study exhibitions, in particular in terms of the reception of narratives by visitors. That is, we need to move beyond the dependence on the visitor surveys mentioned above and instead adopt an interdisciplinary approach using mixed methods. By doing so, we can, so I shall argue, not only acquire the tools to study how such exhibition narratives are received, but also how they are created and mediated. With this in mind, I proposed a research project to explore how Mucem, a museum in Marseille (France) that focuses on the Mediterranean world and its dialogue with Europe, presents narratives in its current exhibition *Connectivities*. As a socially committed museum (*musée de société*), Mucem propagates a multidisciplinary vision, and it is thus a perfect site for studying how complicated narratives are communicated by museums today, and how such narratives—both fed and analyzable by research in anthropology, history, archaeology, art history, and contemporary art—impact on visitors.

Because the content and form of the exhibition and the multimodal nature of the museum's communication is part of the move toward the new types of narratives I identified above, I chose to employ a range of interdisciplinary methods to analyze the exhibition. This enabled me to distinguish my approach from previous practices in museum studies, which, because of the more general tendency toward "evidenced-based" policy and funding in the cultural field, have frequently adopted a positivist approach to their research. Instead, I aimed to analyze not only the intent of the institution when it comes to narrative production but also its legitimacy to and its impact on audiences.

In what follows I will first introduce the Mucem exhibition itself, before outlining the set of methods employed, which encompasses approaches to storytelling in various fields, critical content analysis, and reception through ethnographic research. Next, I will bring these methodological tools together, using them to highlight the study's key findings regarding the disjunction of roles and disciplines in the Mucem exhibition. In so doing, I not only offer a detailed case study of a leading European museum, but I also show that, with a mixed methods approach, we can both analyze the nature of storytelling within the museum and assess whether those narratives translate into meaningful visitor experiences.

## Mucem and the *Connectivities* Exhibition

Mucem, which opened in 2013, has an extensive program of permanent and temporary exhibitions and accompanying public offerings. Because of its prior history, the collection is seen as playing an important role in France's dialogue with North Africa.[1] Recently, Mucem has made an aspirational shift toward wanting to be a global museum, aiming to embed its Mediterranean and European narratives in the histories of the rest of the world.

Within Mucem, I focused my research project on the semi-permanent exhibition *Connectivities*, which opened on June 29, 2020, and runs until March 13, 2023. The exhibition is held in the "Gallerie mediterranée,"

---

[1] The current collection of Mucem is a combination of the collections from Musée des arts et traditions populaire, the European collection of Musée de l'homme, and the collections from the now-abandoned project Musée de l'histoire de France et d'Algérie.

the museum's primary spaces. The exhibition is introduced on the website as follows:

> *Connectivities* tells the story of the great Mediterranean port cities of the 16th and 17th centuries: Istanbul, Algiers, Venice, Genoa, Seville and Lisbon were the strategic sites of power and trade in a Mediterranean that saw the birth of the modern era, between great empires and globalization.
>
> Taking the Mediterranean and the Mediterranean World in the Age of Philip II as its foundation, the exhibition follows in the footsteps of historian Fernand Braudel and approaches this 16th- and 17th-century Mediterranean region not as an object of study with strict chronological limits, but rather as a character with a lengthy story to tell, even extending into the contemporary period.
>
> Inviting visitors to leap backward in time, this urban history continues today, through changes to contemporary port territories like the megalopolises of Istanbul and Cairo and the metropolises of Marseille and Casablanca. This exhibition shows expanding cities as places where influxes, connections trade [sic] and therefore power converge and intensify. (Mucem website, https://www.mucem.org/en/connectivities)

This quote shows the complexity of the narrative proposed and the multiple leaps through time and geography that make this exhibition potentially very difficult for visitors to apprehend.

This narrative complexity is further compounded by the organization of the exhibition, on view in two connected yet distinct spaces. Precisely because of these challenges, the exhibition makes for an excellent case study of the complementary methods for unpacking the exhibition's multimodal narratives, including their impact on visitors. Which narratives are told by the museum, and how does it tell them? How does its audiences perceive and (re)construct those narratives? And how do the narratives presented affect visitors' perceptions of themselves and/or of the museum as a (social) narrative maker?

# Combining Methods: Museology and Its Others

In what follows, I will outline the interdisciplinary methods employed. First, I discuss the use of multimodal storytelling analysis, drawing on the fields of exhibition design, literary studies, and education. Second, I apply critical content analysis, relying on sociology and a subset of discourse analysis. Finally, I develop an approach to visitor reception that builds on ethnography. More than simply mixing methods, however, my aim was to address the exhibition from a perspective that would genuinely integrate those different forms of analysis. Below, I will explain the various methods and the reasons behind their use.

## Multimodal Storytelling

First, I drew on what we might call "multimodal storytelling analysis," employing different strategies to unravel how a narrative is told in the exhibition space. I did so by building on Tina Roppola's exhibition design analysis framework, Mieke Bal's literary analysis tool, Bruce W. Ferguson and Tony Bennett's application of such literary analysis of narratives to museum contexts, and George Hein's model of museum educational theories. The approaches to storytelling outlined in these three fields complement each other and enable me to analyze what objects are shown, how they are shown, who is speaking, what story is being told, how this story is conveyed, and how it impacts the visitor.

In her 2012 book *Designing for the Museum Visitor Experience*, Roppola proposes a framework for analyzing exhibitions design in terms of visitor impact. Roppola distinguishes between four key interconnected design processes: framing, resonating, channeling, and broadening. These processes allow us to account for various types of visitor impact. She acknowledges that these are "interrelated systems," which explains why some elements play out at more than one level (p. 75).

The first of the four processes identified by Roppola, framing, can be considered a "macrolayer." It allows for studying a museum's spatial layout, room(s), and concept organization. The second process, resonating, applies to exhibition displays that "mesh" with the visitor and "achieve some level of kinship" (Roppola, 2012, p. 124), thereby igniting a relationship with the visitor in a short-term interaction. Traditionally, to analyze resonance, the focus is on the visitors' bodily, emotional, and social engagement. The third process, channeling, refers to directedness and cohesion. In Roppola's words, channels are "conduits by which visitors are assisted through the museum, or pathways visitors construct using their own agency" (p. 174). She further distinguishes between spatial, perceptual, and narrative channeling. Finally, the fourth process, broadening, applies to the "content-related meanings" visitors derive from their visit (p. 216). Such broadening may be experiential, affective, conceptual, or discursive in character.

To apply Roppola's framework to the Mucem case study, I collected data by undertaking multiple site observations between February and December 2020. With a particular focus on the abovementioned aspects of Roppola's framework, I used, for making my fieldnotes, forms with sections that encouraged me to consider each of the four relational processes. I also took more free form notes detailing the actual functioning of the exhibition design.

In a second approach, I drew on the seminal work of Bal (1997), who identifies three components of a narrative: text, fabula, and story. She notes that a text can take many forms (book, image, exhibition, etc.), but that, regardless of the form, it always has a narrative structure. Bal defines fabula as "a series of logically and chronologically related events that are caused or experienced by actors" (p. 5). This is the relational, and usually diachronic, aspect of the narrative. Key elements of fabula are events, actors, and time. The final component is the story, which pertains to the manner in which the fabula is communicated, including its ordering, rhythm, use of space, movement, and focalization. These features concerning the story were of particular importance for analyzing the exhibition's wall-texts.

The ways narratives are constructed in museum spaces have also been explored by scholars like Ferguson (1996) and Bennett (1996). Core questions regarding a literary narratological approach to exhibitions are: Who is talking? With which authority? To whom? About what? Once again, I applied these questions to the Mucem exhibition by giving particular attention to wall-texts and labeling, as well as audio and video content.

Third, to complement these design-based and literary narratological approaches to storytelling, I used Hein's classification of exhibition strategies (Sitzia, 2018). In his book *Learning in the Museum* (1998), museum educator and theorist Hein presents a theoretical framework that helps us understand the position of museums when it comes to knowledge and learning. He posits that museums' views on these issues lead to different exhibition strategies. If a museum adheres to a realist view on knowledge, thus considering knowledge as existing independently of the learner, and learning as rather passive and incremental, the exhibition strategy is didactic expository. If the museum approaches knowledge in a realist vein but as actively reconstructed by the learner, the accompanying exhibition strategy is the discovery model. If the museum starts out from a constructivist view on knowledge, assuming that all knowledge is constructed individually or socially, as well as considers learning to be incremental, the corresponding exhibition strategy is the stimulus-response model. Finally, if the museum has a constructivist view on knowledge and assumes that learning is an active process, then the exhibition model will be constructivist. Each model implies a specific strategy of communication and engagement with the visitors, including the choice for and prominence of specific exhibition tools, tone of voice, etc. I used this framework to complement the analysis and identify the museum staff's beliefs and intentions when it comes to knowledge creation and learning.

## Reinforcing Critical Content Analysis with Expert Visits/Interviews

In order to gain critical insight into the exhibition's content—the narrative conveyed—I combined discourse analysis with expert visits and interviews. Discourse analysis enabled me to unveil the meaning implicit in the narrative choices made by the institution. The idea behind the expert visits and interviews is that by visiting the same exhibition with various experts, and by talking about the exhibition with them extensively and critically, the analyst become more aware of the limitations of their own fields of expertise.

In order to protect the experts and ensure that they would feel free to talk, the interviews were anonymous. I selected the following three experts: a curator, with an eye on curatorial expertise of storytelling and the conceptual use of space; an exhibition designer, to assess visitor experience, multimodality, and the use of space; and a historian, to comment on content and clarity. The visits were spread over two months—between June 2, 2020, and October 19, 2020—due to intermittent closure caused by the Covid-19 pandemic. The visits took between 1.5 and 2 hours and were recorded either digitally or in writing, according to the experts' preferences. The material from these visits allowed for reflexive insight in terms of the exhibition's content, in particular in terms of explaining or questioning narrative choices, and what was (or wasn't) in the exhibition. It allowed for a more refined exploration of the choices made by the institutions in terms of what story to tell and, to some extent at least, the reception of these narratives.

## Narrative Reception: Ethnographic Observation and Interviews

The third kind of method employed involved investigating the reception of the narrative by audiences using ethnographic observations of museum visitors as well as exit interviews with visitors. This gave me insights into how people were behaving in the exhibition space, to establish how the narrative was being read by the audience, and to evaluate the impact of the institutional storytelling choices through various forms of visitor engagement. Visitor research is often hailed as the only way to truly evaluate the impact of an exhibition. Indeed, an exhibition can work "in theory" and yet be completely misinterpreted by the audience.

Ethnographic observation of visitors allowed me to look at how people were moving around the space, to establish their paths, to investigate what visitors were reading (or not), and what they were looking at (for how long and in what way); it also allowed me to listen to them while they exchanged views on the content of the exhibition (Walsh, 2012;

MacDonald, 2010). I did the ethnographic observations between June and November 2020, with visits of varying duration, observing a variety of visitors in terms of age and socio-cultural background. I focused on aspects that were suggested by the primary space analysis: I studied visitor flow, their orientation and movement in space, their reading of and engagement with the written material, their engagement with multi-modal forms of discourse, and the relationship they created with the objects.

I purposefully opted for "quick-fire," short-form exit interviews, undertaking 45 of them in total, and asking just one single question: "What is the main message/idea you take from the exhibition?" The interviews were conducted in each visitor's mother tongue (i.e., mostly in French, except for three interviews in English and one in Dutch). Next, I coded the interviews and analyzed them thematically.

Finally, to complete my data, I conducted interviews with the two exhibition curators (other than the curator selected for the expert visit) to gain insight into institutional decision processes and help the institution rethink its narrative creation and exhibition process. The semi-structured interviews, which lasted about an hour for each curator, were also essential in building a constructive relationship with the institution.

Overall, the interdisciplinary approach outlined here allowed me to gather the necessary data about various facets of exhibition storytelling, about institutions as active makers of social narratives, and about the impact of such narratives on visitors. I will now outline the main findings of my case study, with a particular emphasis on the intersectionality of the methods used.

## Main Findings: At the Crossroads of Disciplines

My research identified multiple issues within this exhibition, including Eurocentrism and the disappearance of contested history, as well as the issue of sensory overload (Sitzia, 2022). While these are both fruitful and important areas for future study, for the purposes of this chapter I would like to focus on one finding in particular: the disjunction between the roles the museum assigned itself and the discourses it conveyed in its space. In doing so, I will also demonstrate the various ways in which the mixed method approach enabled a more nuanced and elaborate reading of the exhibition's narratives.

## Tensions Between Exhibition Models

A first symptom of the disjunction of roles and discourses was revealed through the use of storytelling analysis tools and frameworks. Specifically, this could be seen in the tension within the exhibition between two of Hein's exhibition models: the didactic/expository model and the constructivist model. Didactic/expository models are usually connected to a perception of the museum as a traditional educator, as a transmitter of information and a holder of knowledge and authority. Constructivist models are connected to the perception of the institution as a place of reflection and debate—a vision of the museum as a public forum.

By combining Hein's educational models with Roppola's exhibition design framework, we can see that *Connectivities* is framed as a "spectacular" exhibition. Exhibition design choices—such as the lighting (especially in the contemporary part of the exhibition), the way objects are presented in an aestheticizing manner, and the sound level—place the exhibition in an expository logic. The topic of the exhibition itself—framed by an established, relatively old-fashioned, and complex academic framework such as Braudel's[2]—firmly places the institution as a displayer of ideas and the exhibition as a didactic experience.

However, the room organization of *Connectivities*, which is firmly constructivist, actually contradicts this didactic/expository position. The exhibition offers a free path, while the double linear narrative—one following the sixteenth- to seventeenth-century narrative of the Mediterranean and one looking at contemporary Mediterranean urbanism—is interrupted with regular openings between the various spaces.

---

[2] Historian Fernand Braudel (1902–1985), an advocate of historical materialism, is well known for his "longue durée" perspective on history that considers social, economic, and cultural dimensions as closely interconnected. His work on and approach to the sixteenth- and seventeenth-century Mediterranean region—as a multitude of exchanges rather than an object of study with strict geographical and chronological boundaries—constituted the starting point for the exhibition.

Here, the storytelling analysis was backed up by my ethnographic observations, which showed the impact of this tension on visitors. Indeed, a majority of visitors were at first looking for information throughout the space; they read the labels and wall-texts carefully (when provided), yet most visitors looked fruitlessly for extra information. Visitors alternated this search for information with contemplative moments in relation to the objects on display.

The critical content analysis of the videos also confirmed this tension. The historian expert noted that the introductory Braudel video is not an introduction to Braudel's work, but rather a presentation of comments on the impact of Braudel's work by experts in various fields. It is a patchwork video, which is the kind of format one would expect in an advanced constructivist context where various points of view are presented to let visitors develop their own position. As an entry point to a didactic/expository exhibition, this video makes it difficult for anyone unfamiliar with Braudel to understand what the exhibition is about. Nor was it clear to visitors, as I established, that this video creates a link between the two paths (according to the exhibition curator).

Interestingly, the storytelling analysis showed that the exhibition has an educational and highly didactic label explaining Braudel's theory, but that this label was located a few meters away from the video introduction. The ethnographic observation showed that this aspect of the tension between didactic/expository and constructivist exhibition codes disoriented audience members as soon as they entered the space. Most visitors stopped briefly in front of the patchwork introductory video, moved on rapidly to the exhibition, returned to the screen again and again, trying to connect the para-discourse on Braudel to the objects on display, looking for a red thread and often missing the description of Braudel's theory label. Furthermore, the exhibition design expert pointed out that there is a wall-text introducing the overall argument of the exhibition but that it is badly placed (close to the entry door on the right when entering) and that this label also has very little visibility (it is under-lit and in a small font). This is confirmed by my ethnographic observations, as I saw only a tiny minority of visitors (3 out of 132 in total) who read it.

Here, then, the interdisciplinarity of my methods did not only allow me to identify an issue but also to explain it in nuanced ways. This approach, moreover, had a practical implication for Mucem: it gave rise to my recommendation to reorganize the introductory space.

## The Use and Presentation of Objects

The tension between exhibition models is also visible in the way objects are presented. Looking again at Roppola's process of framing, the materiality of the exhibition reinforces the impression of spectacle: objects were lit dramatically, contained in glass boxes, and often without labels in proximity. This exhibition design analysis was confirmed by the expert curator who identified an issue with the register of presentation, noting that all objects were presented at the same level (maps, artifacts, models, artworks, etc.). This confused the status of the objects as documents or monuments; that is, it encouraged visitors to read all objects as documents or "clues," while also presenting them as artworks. This contradiction led to the hesitant visitor postures that I identified during the ethnographic observation; that is, their behavior read somewhere between information seeking and contemplative admiration.

The expert scenographer formulated a similar concern, highlighting "the domination of objects" in the space. When probed, the scenographer insisted that the aestheticizing presentation of objects (behind glass, on pedestals, etc.) conveys a sentiment of exclusivity, which is especially problematic as several objects are emptied of their message and mediation tools are excluded or marginalized. For example, a large case of china was presented without reference numbers, making it impossible for viewers to link the content of the labels to the pieces on display. This also explains the visitor uncertainty identified in the ethnographic observation: some visitors had difficulties identifying the objects and placing them in the broader narrative of the exhibition and so they circled around the artifacts and looked for specific information related to them (often without success), while other visitors, when in front of the objects, behaved as if these objects were artworks.

It is here that the importance of the combination of methodologies from different disciplines becomes evident, as Roppola's concept of framing and the expert contributions of the scenographer allow us to better interpret the results of the visitor exit interviews. In particular, during the quick-fire interviews, three main categories of interpretations of the exhibition narrative by the visitors emerged: (1) their reappropriation of the narration, in particular concerning specific cities, in terms of the familiar, triggering recognition and reassurance; (2) their use of very general concepts such as "the Mediterranean," "urbanism," and "diversity"; and (3) their focus on specific objects, such as a boat model, tile, painting, or coat of arms. It is this third category that is well explained by the storytelling and content analysis above and thus by the interdisciplinary mix of methods.

## The Tone of the Narration

The disjunction of roles and discourses is also visible in the tension between the various tones of the narration, which we can analyze through Roppola's framework. Put bluntly, the experiential broadening proved a jumble because it failed to offer a coherent experience to the visitor. The narration of conflicted relationships in the Mediterranean in the contemporary section contrast with the presentation of polished relationships in the historical section, creating an affective disjunction between the two parts. In addition, the conceptual broadening of the narrative gives priority to urbanism without clearly delineating this notion. This plays out in the significant number of visitors who focused on particular cities when asked about the main message of the exhibition. The discursive broadening is all the more an issue because the texts are very directive and didactic in tone and leave little room for individual reflection.

Furthermore, this tension can lead to critical misinterpretation by the audience, as the interviews show. For example, one visitor said to be astonished about how well nations got along in the sixteenth century. Not only did such tensions obviously impact visitors' historical understanding, they also resonated in examples of the exhibition's Eurocentrism and omissions in terms of postcolonial perspectives. It is worth briefly noting some of these instances: several North African cities (Algiers, Tripoli, and Tunis) were combined in a single label, while each European city was afforded its own label. The expert curator also observed that while Soliman the Magnificent is the subject of one particular label and its related objects in a specific exhibition section, the portrait of François I dominates the wall. Furthermore, the expert historian noted that the use of terms such as "occidental civilization" in some labels is problematic in a postcolonial reading of Mediterranean history. Finally, the expert scenographer and historian conceded that 200 years of history are missing (without justification) and that there is no explicit mention of colonization, with the expert curator also expressing surprise: "Nothing on slavery?!" In contrast, a strategy clearly distinguishing between multiple voices and intersecting perspectives would have permitted a more balanced discourse.

By further combining our methods of analysis, we can actually gain a better understanding of the ways in which the curators' aims fail to play out in the exhibition. Returning to Hein, we can see that the exhibition curators intended to follow a constructivist approach, as they say the exhibition is trying to trigger a "personal and social engagement with cities and connections" (interview with curator 1). However, the critical content analysis revealed that the tone in the labels is that of a demonstration. Furthermore, the expert curator noted obscure expressions—as seen in the Istanbul label's inclusion of "cultural syncretism"—which do not suit a general audience. In fact, not all objects were labeled, even though one of the arguments of the exhibition aimed to promote was the "circulation of objects" (interview with curator 1). The expert curator expressed surprise in this regard, stating the need for more explicit object labeling, as in the curator's view "they don't speak by themselves."

Another example of this issue of tone and register is the timeline on architecture, which is very difficult to read for non-experts. The double discourse design (with parallel top lines for Europe and bottom lines for North Africa), its text heaviness, and the use of expert architectural vocabulary make it difficult to access. This analysis was confirmed by the historian expert, who mentioned that conceptually complexity of this exhibition element, as it tries to outline issues of architectural cross-fertilization without mentioning orientalism or colonialism explicitly. The ethnographic observation confirmed that only a few expert readers were at ease; rather, a large majority abandoned attempts at reading after a couple of minutes. The problem is, however, that the timeline presents the conceptual grid through which to read the rest of the contemporary path. This leads to the visitors' "city" focus devoid of the "connection" argument, as testified by the quick-fire interviews.

The different methods of analysis were also in line with each other in terms of the findings concerning the exhibition's entrance. The expert curator, for instance, highlighted that there is no buffer space at the exhibition entry, which instead opens with the patchwork video presenting the comments on the impact of Braudel's work. Traditionally, this entrance space would be used to clarify the intention of the curators. Ethnographic observation confirmed that the disorientation at the beginning was carried on through the exhibition, and even amplified by the fact that the map at the entry does not match the actual space but rather creates a symbolic image of the Mediterranean—something mentioned by one of the curators but identified by none of the experts or visitors. This created hesitation and disorientation in almost all visitors.

This disorientation was aggravated by the exhibition's use of two paths: a historical and a contemporary route through the exhibition. As established through ethnographic observation, some visitors tried to follow one of these paths but ended up stuck at the end and had to go back to the entry to start with the other path. Alternatively, they followed the other path in backward order, losing its narrative structure. My observations also revealed that some visitors switched between the historical and the contemporary paths, using the open spaces to move from one path to the other. These openings were meant as "windows between the spaces" (interview with curator 1), but at times this completely disoriented visitors. Furthermore, most people had trouble finding the exit—hidden as it was behind a large screen. The display of the exhibition sponsor's video close to the exit contributed to this confusion, as it gives one the impression of being a conclusion, a summary of both paths.

## Conclusion

From this layered interdisciplinary analysis, we can conclude that the institutions and the curators need to make clearer choices for their exhibition: that is, as either a didactic/expository or as a constructivist space, and in terms of their use of objects as well as their narrative tone. Not making such choices creates confusion among visitors and can lead to misunderstanding of the exhibition's argument. Ensuring that both the entry and advanced levels of the information are consistent, and making the signaling clearer would go a long way in solving these issues. Collaborating with focus groups and linking the exhibition more closely to today's world might help to counter the issue of Eurocentrism.

The combination of methods from various fields allowed me to generate a sharper and more detailed analysis of storytelling processes in exhibition spaces as well as of the reception of such narratives by the public. My mix of methods proved efficient in particular for analyzing complicated multimodal environments. It also offered a more layered explanation of the results and provided a better understanding of causality, especially when it came to certain visitor's interpretations. It thus helped the research to go beyond the traditional conclusion that "it doesn't work."

Indeed, it is the integration of methods that allows for a rich analysis adaptive to the dynamic landscape and inherent complexity of museum narratives with multiple enunciators, receptors, and modes of communication. This also helps to unpack the institution's assumptions and, in turn, to contribute to transforming the field. Museums should be able to present complex, rich, and multivocal narratives. They should invite visitors to wander and wonder, but without causing them to get lost in the exhibition space.

# References

Antos, Z., Fromm, A. B., & Golding, V. (Eds.). (2017). *Museums and innovations*. Cambridge Scholars Publishing.

Bal, M. (1997). *Narratology: Introduction to the theory of narrative*. University of Toronto Press.

Bennett, T. (1996). The exhibitionary complex. In R. Greenberg, B. Ferguson, & S. Nairne (Eds.), *Thinking about exhibitions* (pp. 81–112). Routledge.

Borg, C., & Mayo, P. (2010). Museums: Adult education as cultural politics. *New Directions for Adult and Continuing Education, 127*, 35–44. https://doi.org/10.1002/ACE.379

Chynoweth, A., Lynch, B., Petersen, K., & Smed, S. (Eds.). (2020). *Museums and social change: Challenging the unhelpful museum*. Routledge.

Clover, D. E. (2015). Adult education for social and environmental change in contemporary public art galleries and museums in Canada, Scotland and England. *International Journal of Lifelong Education, 34*(3), 300–315. https://doi.org/10.1080/02601370.2014.993731

Davallon, J. (1999). *L'exposition à l'œuvre: Strategies de communication et médiation symbolique* [The exhibition at work: strategies of communication and of symbolic mediation]. L'Harmattan.

Dewey, J. (2008). *Democracy and education.* Project Guttenberg. https://www.gutenberg.org/files/852/852-h/852-h.htm#link2HCH0008 (Original work published 1916)

Ferguson, B. (1996). Exhibition rhetorics: Material speech and utter sense. In R. Greenberg, B. Ferguson, & S. Nairne (Eds.), *Thinking about exhibitions* (pp. 126–136). Routledge.

Hein, G. (1998). *Learning in the museum.* Routledge.

Janes, R. R., & Sandell, R. (Eds.). (2019). *Museum activism.* Routledge.

MacDonald, S. (Ed.). (2010). *A companion to museum studies.* Wiley-Blackwell.

Mairesse, F., & Desvallées, A. (2007). *Vers une redéfinition du musée* [Towards a new definition of museums]. L'Harmattan.

Marstine, J. (2006). *New museum theory and practice.* Blackwell.

McSweeney, K., & Kavanagh, J. (Eds.). (2016). *Museum participation: New directions for audience participation.* MuseumsEtc.

Mouffe, C. (2016). An agonistic conception of the museum. In *Proyectomuseu,* n.p. http://proyectomuseu.org/an-agonistic-conception-of-the-museum/

Roppola, T. (2012). *Designing for the museum visitor experience.* Routledge.

Sandell, R. (1998). Museums as agents of social inclusion. *Museum Management and Curatorship, 17*(4), 401–418. https://doi.org/10.1080/09647779800401704

Simon, N. (2010). *The participatory museum.* Museum 2.0.

Sitzia, E. (2017). The ignorant art museum: Beyond meaning-making. *International Journal of Lifelong Education, 37*(1), 73–87. https://doi.org/10.1080/02601370.2017.1373710

Sitzia, E. (2018). The many faces of knowledge production in art museums: An exploration of exhibition strategies. *Muséologie: Les cahiers d'études Supérieures, 8*(2), 141–157. https://doi.org/10.7202/1050765ar

Sitzia, E. (2022). Senses and sensibility: Finding the balance in sensory museum education. In A. Sinner (Ed.), *Artful xchanges: Propositions in museum education (n.p.).* Intellect Books. (forthcoming).

Vergo, P. (1989). *The new museology.* Reaktion Books.

Walsh, D. (2012). Doing ethnography. In C. Seale (Ed.), *Researching society and culture* (pp. 245–262). SAGE.

# Spatial Rituals and Ritualized Space in Dutch Postwar Homes for the Elderly: Anthropology in History

Karin Bijsterveld

## Introduction

In the late 1980s, the Indian anthropologist Sanjib Datta Chowdhury spent time in the Netherlands conducting ethnographic research as a student nurse in a residential home for the elderly. He noticed that its elderly Dutch residents inhabited these homes in ways entirely unfamiliar to him. They would gather together in the home's recreation room and coffee corners but never visited each other's studio apartments (Chowdhury, 1990, 1995). At first, Chowdhury could not make sense of these codes. It was only after examining the spatial layout of rooms in typical single-family Dutch homes, and these rooms' designations as either public or private spaces, that he understood the spatial rituals of residential homes.

K. Bijsterveld (✉)
Department of Society Studies, Faculty of Arts & Social Sciences,
Maastricht University, Maastricht, The Netherlands
e-mail: k.bijsterveld@maastrichtuniversity.nl

K. Bijsterveld, A. Swinnen (eds.), *Interdisciplinarity in the Scholarly Life Cycle*,
https://doi.org/10.1007/978-3-031-11108-2_10

When Chowdhury published his research, I was writing a postwar history of Dutch homes for the elderly. A historian by training, my specialization at that time was policy analysis. Initially, my narrative entailed a rather traditional overview of national policies concerning these residential homes, largely in the actor terms of policymakers that highlighted good care and independence as key notions. Chowdhury's spatial-anthropological approach dramatically changed both the type of sources I researched and my analysis. First, I began to more closely examine the architectural plans and photographs of residential homes in policy reports and other publications. Second, the ways in which these plans and photos marked spaces as private, public, or semi-public enabled me to question how middle-class ideas about how to occupy space inspired the layout of residential homes.

This chapter shows how my anthropology-in-history developed and what this contributed to Chowdhury's interpretations of *spatial rituals*. At the same time, it offers a more performative reading of this architecture and its representation in plans and photos, as *ritualized space,* might further enrich the analysis and deepen interdisciplinary integration.

## Spatial Rituals: An Anthropologist's Take on Homes for the Elderly

Raised in India, where older parents typically lived with their children and grandchildren, Chowdhury expected that the elderly people in the Dutch residential home he was studying would be lonely and unhappy. Sociologist Erving Goffman's work on life in total institutions such as psychiatry wards also informed Chowdhury's assumptions, most likely because Goffman considered homes for "the aged" as a subtype of total institutions (1961/1991, p. 16). With that in mind, Chowdhury initially looked for the "mortification of the self" and the existence of an "underlife" of inhabitants among each other (Chowdhury, 1990, p. 32).

To Chowdhury's surprise, he found markedly fewer signs of these phenomena than anticipated. Instead, he made detailed field notes about what he observed to be spatially bound interactions among residents and between them and visitors from the outside world. Although inhabitants of the residential home refrained from meeting in their studio apartments, they *did* receive family members there, but only by appointment. They often announced such family visits emphatically at the recreation room table; however, the visits themselves remained out of view of other residents. Residents would occasionally go out together or take care of each other's plants, mail, or groceries in cases of absence or illness. Still, it did not change anything fundamental in their attitude toward visiting private apartments (Chowdhury, 1990, 1995).

Chowdhury also observed the ways that the elderly met each other outside their studios. Women considered it of great importance to be carefully dressed when coming to the recreation room for coffee: the anthropologist noted the use of the *bloemetjesjurk* (flower dress) for these occasions. In the recreation room, most inhabitants had their own fixed place at a particular table, where they met their friends and conversed with one another. It was not acceptable, though, to complain about one's personal health. Rather, it was important that the conversation remain light and *gezellig* (cozy). The residents particularly liked chatting about "supermarket prizes, weekend outings, and visits from children" (Chowdhury, 1995, p. 150, translation KB). Flirting with caregivers and discussing the horrors of the meals served by the kitchen were also widely practiced.

What do we make of all these spatial rituals—of these codes of inclusion and exclusion? Chowdhury did not fully understand how they worked until he began to scrutinize the notion of privacy in the use of space in typical middle-class, single-family Dutch homes. As he explained, these homes have hallways guaranteeing that all rooms are independently accessible. This architectural feature allows household members to meet "external" visitors in distinct ways. If the visitor is someone known to the entire family, the guest will be welcomed into the "public" living room. In case the visitor is an intimate friend of only one of the family members, especially if this concerns one of the children, this friend will be received in the relevant member's "own" private room.

This way of distinguishing social relationships through the use of space extended to residential homes, according to Chowdhury. In these homes, the inhabitants' studio apartments functioned just like the individual rooms of single-family homes: as private spaces. The other spaces, such as the hallway, coffee corner, recreation room, or library, acquired the meaning and purpose of public areas. Unlike the single-family home, the individual studios in the residential homes did not offer the possibility to differentiate social relationships. Inhabitants thus transposed the act of differentiation to other spaces, like meeting friends at the recreation room table or briefly lingering with other residents in the hallway. This was their way to implement autonomy in their companionship with others. They still valued their privacy but associated it with alternative spaces and settings. As living in a single-family home was considered an important achievement for the elderly under study—many of them had experienced much poorer, one-room housing conditions during the prewar era—they treasured this division of space. If a resident were to become less mobile over time, these spatial rituals might result in isolation and loneliness, as Chowdhury learned through interviews. Yet this was the price the elderly paid for their culturally informed designation of particular spaces as "private" (Chowdhury, 1995, p. 159). To situate this theoretically, Chowdhury returned to Goffman's work—this time, to his notion of the "presentation of self in everyday life" and the differences between those presentations in distinct situations (Goffman, 1959/1987).

I found Chowdhury's interpretation highly convincing when I first learned of his research. Although difficult to admit today, I had never read work by a non-Western anthropologist writing about Western institutions before the late 1980s. I was familiar with publications by Western scholars on non-Western cultures, which significantly informed my relativist understanding of knowledge. My exposure to Chowdhury's work, then, was a truly eye-opening experience. It not only made me aware of my complicity with the default neocolonialist attitude of viewing non-Western culture from a Western perspective, but it also inspired me to revisit the ways in which I had previously approached debates about aging populations and housing the elderly.

# Living Independently: A History of Discourse on Housing Aging Populations

By the end of the 1980s, nearly every municipality in the Netherlands—village, town, or city—offered a residential home for the elderly with annexed dwellings. These dwellings were small, terraced houses serviced from a main building: the residential home. The home itself commonly had four to eight floors with long corridors, modest one- or two-room apartments, balconies of a few square meters, and orange-colored blinds. Today, one can still find these buildings scattered across the country, some still homes for the elderly. Yet many of the previous homes have been demolished, turned into expensive luxury housing for "seniors" and not-quite-"seniors," or are patiently waiting for new designations.

What happened in between was the rise of a national policy discourse promoting the importance of independent living for the elderly. Initially, it was my idea to trace the origins of this idea by analyzing national policy reports, parliamentary debates, and expert publications on residential housing and situating these against the background of the rapidly developing welfare state. However, after finding Chowdhury's article in my office mailbox—placed there by a colleague from the philosophy department—I had to acknowledge that the narratives emanating from the reports invited additional research. I had to do more than merely write a contextualized history of ideas, and I needed to examine architectural plans and photography in addition to the policy reports.

While my initial argument identified a radical shift in Dutch national policies concerning the elderly, from fostering residential housing to independent living—a shift that had already started in the early 1970s and was in full swing by the time Chowdhury conducted his research—my aims changed after reading his ethnographic work. First, in addition to clarifying the shift away from residential housing, I also wanted to understand the rise of the modern old-age home in the first place. Second, I now aimed to unravel how widely embraced postwar residential home policies responded to what these buildings replaced *and* to what extent such policies already drew on a rhetoric of independence. Until reading Chowdhury, I had either barely noticed or simply dismissed as less relevant references to the notion of independence in the pre-1970s era. I now realized that these should be part and parcel of my argument and that I should write a conceptual-material history, showing that "independent living" had been an ideal defended since the early postwar years, but that its meaning and materiality had radically changed over time. Chowdhury's socio-anthropological analysis of the cultural connections between material spaces across private-public axes and distinguishing between different social relationships worked as levers for opening up my own analysis (Bijsterveld, 1996).

National policy reports and expert publications on elderly care published in the first half of the 1970s provided the outline for the following "crash-course history" of old age and housing. Once upon a time, elderly people died much younger than they do today and were cared for until then by children and other family members living with them. After World War II, however, the Dutch state began to acknowledge the rapid aging of the population and worried about how to care for the growing demographic of people aged 65+ in an increasingly individualistic world. These conditions fostered the establishment of large-scale residential housing for "old" people, as they were still considered at the time. The country's daunting postwar housing shortage also played a role, as the single-family homes vacated by the elderly would become available to the remaining population. The introduction of a pension in 1957 for all those 65 and older (de *Algemene Ouderdomswet*) enabled the rise of residential housing by providing the elderly the means to move to these homes. The introduction of the *Algemene Wet Bijzondere Zorg* (AWBZ) or General Law for Specialized Care in 1965 facilitated this even further. While in 1950 only 3.8 percent of people aged 65+ lived in residential homes, this figure had increased to 8.9 percent by 1975 (Bijsterveld, 1996, p. 208). In 1950, the Netherlands had 812 residential homes, a number that grew to 1880 by 1970 (Bijsterveld, 1996, p. 209).

Around 1970, however, the tone of the policy documents changed. In the context of deteriorating economic prospects, policymakers considered the costs of these facilities too high; increased costs resulted from the rapidly increasing average age of residential home inhabitants as well as rising staff salaries. Policymakers also acknowledged that it had not been such a good idea after all to accommodate the elderly in large institutions, thus leaving them inactive. Instead, the elderly should live as independently as possible for as long as they could. This would keep them mobile, benefitting both their health and happiness. The key notions in thinking about housing for the elderly thus shifted from dependence to independence, from intramural to extramural, and from segregation to integration into social life (*Nota Bejaardenbeleid*, 1970, 1975).

These historical overviews in policy reports were rather Whiggish, however. They assumed that the idea that seniors could take care of themselves and "live independently" was a novel invention or ideal identified by policymakers. As I myself initially believed, following this account uncritically, the dominant narrative about residential homes was that they had been meant to take care of the elderly entirely, making seniors inactive and dependent as a result.

## Ritualized Space: Residential Homes as Family Homes, Hotels, Villages, and Suburbs

However, while the building of residential homes was indeed put to a halt as of the 1970s, the idea that the elderly live as independently as possible for as long as possible was not new at all. The only thing that changed, in my view, was the materialization of that idea. As stated above, it was Chowdhury's attention to space that inspired me to study the initial plans and architectural layout of residential homes in more depth, as well as the way in which these plans and their accompanying texts designated different spaces as private, public, or semi-public.

My alternative history began with the idea that residential housing for the elderly actually had quite a long history. As early as the late medieval era, unmarried women or childless widows, usually poor, could find shelter in *hofjes* of their respective religious denominations. In the nineteenth century, foundations for poverty relief and philanthropy, both municipal and private ones linked to churches, established asylums (*gestichten*) for the "invalidated" (including people of old age) on a larger scale. By the late nineteenth century, the Netherlands also witnessed the rise of commercial homes for the elderly, as their number had exceeded what religious foundations could accommodate, notably in urban environments. This was further stimulated by the establishment of modest old-age pensions for the poor and "invalidated" (*Invaliditeitspensioen*). Like the asylums, these homes commonly offered one room for men and one for women. Each also had a "father" and a "mother"—usually a couple—whose official duty was to care for the elderly, although they often asked for as much rent as possible in exchange for as little care—sometimes authoritarian and denigrating in nature—as possible. The material conditions of these homes were often quite dreadful as well.

By the 1910s and 1920s, some philanthropic institutions noted that the elderly population in need of care was changing in character. Some who had once been "better situated" or "middle-class" were impoverished due to interwar inflation. Such people did not really belong in asylums or commercial homes. The institutions, then, fostered the idea of *classificeering* (classification) and established a new type of guest or boarding house for the "civilized" or "dignified" classes. These homes offered housing for some 80–140 elderly residents, were often backed by churches, and offered sliding scale fees. Some inhabitants paid a fee while others received support from church *diaconie* (poor relief departments). The higher the class, the higher the prices and the more privacy one would be entitled to: less elderly people per room, distinct *chambrettes* (alcoves in large bedrooms), or a room of one's own. Competition between religious denominations also inspired the establishment of these homes (Bijsterveld, 1996, p. 162). The basic idea, however, was to couple class and income with levels of privacy enacted through space.

Some of these guesthouses were rather small in scale (25–30 inhabitants) and even situated in richly decorated villas surrounded by lush, park-like settings. As they were meant to function more like hotels, these upper-middle-class guesthouses were known as *pension-tehuizen*. The foundations (Pro Senectute, Vredeheim, Ons Thuis) that funded these homes were explicitly against commercial exploitation and aimed to offer a more traditionally home-like environment with a less patronizing tone than their poorer equivalents. Additionally, the preservation of the elderly's "untouched independence" was a key goal. Making these homes homelike was similarly important—for instance, by providing a *gezellig zitje* (cozy corner) or a sun lounge, as I have underlined in my analysis of photos and captions published in venues by and for social care professionals (Bijsterveld, 1996, pp. 210–16).

These guesthouses set the example for the rise of residential homes for the aged (*bejaardentehuizen*) in the 1950s and 1960s. After World War II, the Dutch government emphasized that housing for the aged had entered a new phase. No longer was it a form of philanthropy but should be seen as specialized housing for those who were entitled to it given their age and health conditions. This specialized housing could remedy some of the housing shortage but only to a limited extent, as most elderly people would leave bad-quality housing, abandon commercial homes, or end cohabitation. Like guesthouses, the new residential houses were designed to both preserve their inhabitants' independence and freedom *and* to provide service, sociability, and community life. This independence and freedom combined with service and companionship were put forward to contrast with the lack of freedom and private space in the earlier asylums and commercial homes. With aims thus focusing on both the individuality and collectiveness of the elderly, albeit with variations in degree, the private foundations and corporations behind the homes began fostering conceptions of the homes as family homes, hotels, villages, and suburbs. This happened both explicitly, through discourses on housing, and more implicitly, as I will show below, through photography and architectural floor plans.

Initially, national authorities expressed a preference for a clear choice between either small- or large-scale guesthouses. The small ones, with about 30 residents, could still preserve the character of a big *family* and would remain embedded in a neighborhood. Larger homes, with 100–300 inhabitants, had to abandon the notion of a family and were better conceptualized as *hotels*. In practice, however, economic considerations inspired a trend toward grander homes with more provisions. Interestingly, Catholic and Protestant foundations expressed different ideas about what they should offer. The first highlighted the importance of creating a genuine, *village*-like community that would enable the elderly to remain there until their death. This implied that the residential home should include a hospital within its walls. In contrast, Protestant organizations approached residential homes as *suburbs*. The homes should function more like neighborhoods, where residents could keep their own general practitioners and be able to go to hospitals of their choice. Architects subscribing to this suburb version went so far as to present floor plans of homes featuring studio apartments with separate entrances along street-like hallways, whereas the village option came with front doors that opened onto a circular or distinctly colored space designed to evoke a central village square (Bijsterveld, 1996, pp. 217–20).

Nearly everywhere, however, the residential homes grew bigger in the 1960s and 1970s, often featuring granny flats as well. By providing rooms with private entrances, nameplates, letterboxes, and doorbells, the rooms developed into small apartments (Bijsterveld, 1996, pp. 187–88). To underline the independence of the apartments, their doors opened onto open-air corridors that grew longer and longer, due to the sheer number of rooms. Through "smart furnishing," rooms could include living and dining areas and even a piano (Anonymous, 1955, as cited in Bijsterveld, 1996, p. 218). In addition, this suburb might have a (cigar) shop, hairdresser, billiard table room, recreation room, theater, television room, library, telephone booth, church, and mortuary. The entrance of the mortuary should be out of sight, although one director of a residential home noted that watching funerals was often the "event of the day" for the elderly (Rubbens-Franken, 1957, as cited in Bijsterveld, 1996, p. 178).

In line with both the village and suburb approach was that the relationship with the world outside the residential home had to be that of "sheltered connection." The elderly should be able to experience city life without enduring any of its nuisances. For that reason, high-rise buildings were considered most appropriate. Garden design should focus on use by the elderly, that is, to be looked at more than to be walked through: "Sitting in the garden will be rare, unless special wind screens of glass make this very appealing on a nice day" (Anonymous, 1949, as cited in Bijsterveld, 1996, p. 186). While creating combinations of shelter and openness was relatively easy for the architects to do by applying glass and high-rise constructions, creating both privacy and a sense of community was more challenging. After all, the massive size of the buildings made it hard to express coziness or togetherness. Architects, therefore, tried to soften the monumentality of their designs, for example, by emphasizing horizontality or by offering inhabitants a pleasant lobby (Bijsterveld, 1996, p. 186).

Whether seen as family homes, hotels, villages, or suburbs, connecting the notion of independence to a range of spatial solutions dominated the discourse on residential housing during the entire postwar era until the moment that the homes lost their appeal to policymakers. Thereafter, residential homes were only offered to a limited percentage of the elderly—a maximum of 7 percent, which was further reduced over time. By the mid-1970s, however, residential elderly homes already dominated the landscapes of Dutch villages, towns, and cities, inspired by and materializing in a spatial representation that I would like to call "ritualized space."

# Postcards and Architectural Photography

Revisiting my interdisciplinary approach today, however, I do not think I traveled far enough down the anthropological road. That I only went so far is partially understandable: we cannot project ethnographic research back in time. The most obvious alternative, an ethnographically informed oral history project, could not materialize because most of the relevant elderly people were deceased by the time I developed an interest in their experiences from the 1950s and 1960s. Interviews with staff or former staff of residential homes could have been an option (Greubels, 2020) but would not have resulted in a first-person perspective of how inhabitants enacted or experienced spatial rituals and ritualized space.

Something close to a historical ethnography of residential homes would have been possible with earlier published prosopographies, that is, group biographies or ethnographies of residential home inhabitants. In the mid-1960s, a participant observation study of six Dutch residential homes focused on their ideal size but was unable to provide conclusive advice in this respect (Nierstrasz, 1965). Recently, there was a group por-trait of residential home inhabitants albeit in the form of a fictional narrative, penned under the pseudonym Hendrik Groen (Groen, 2014/2018, 2018/2020, for a critical reading, see Swinnen, 2019) and turned into a highly popular television series (Oliehoek, 2017–2019). At the time, however, I did not have access to such a source, nor to the kinds of questionnaires that proved highly useful in understanding previous experiences of unmarried elderly women (Bijsterveld et al., 1992, 2000) or drivers' experiences of highway noise barriers (Bijsterveld et al., 2014).

I had entirely overlooked a wonderful source, however. Once again, two writers with educational backgrounds very different from my own brought this source to my attention. In 2019, musician and designer Sonja van Hamel and graphic designer Robert Musa published *We mogen niet klagen: Kaarten uit het bejaardentehuis* (We Shouldn't Complain: Postcards from Residential Homes). The authors collected hundreds of postcards sent in the 1970s, 1980s, and 1990s and featured 138 of them, posted from all over the Netherlands, in their book (Van Hamel & Muda, 2019). They likely did not select these postcards in a systematic or representative manner; after all, they were not historians but professionals with an eye for the design of the buildings and the postcards themselves.

Nevertheless, the postcards provide a very rich source for historical study. In retrospect, they enabled me to reinterpret Chowdhury's conclusions as well as my own. The texts on the postcards represent first-person narratives written and mailed by residents to friends, former neighbors, and family members. At the same time, the postcards' images captured the homes' design in optimal forma, showing the residential homes' preferred self-image in their most colorful version, bathed in sunlight, with all or nearly all the orange-red blinds—along with some green or blue ones—drawn.

Studying this published collection, Goffman's work on the *presentation* rather than the *mortification* of the self appears to me just as relevant as it was for Chowdury's eventual analysis of residential life, but with a twist. Goffman analyzed everyday interaction in dramaturgical terms, showing differences between the front *region,* or front stage, and backstage communication—think of differences in communication between employees of a car repair business on the shop floor (backstage) versus employees and customers in the shop (front stage). He was particularly interested in the precautionary work invested in preventing disruptions of how teams present themselves at the front. Rather than interpreting front-stage behavior as superficial and back stage as authentic, he considered both sides real and key to the social fabric of life. He also saw the use of front-stage language as the "absence (and in some sense the opposite)" of what was informally accepted backstage language (Goffmann, 1959/1987, p. 129).

Some of the writing on the postcards reflected the behavior that Chowdhury had identified as appropriate table conversation among residents—"we shouldn't complain"—thus preserving front-stage behavior. Yet most texts expressed an ambivalent mixture of pride, tentative adaptation (to perhaps reassure those left behind), and a somewhat concealed desire for company. "I have already begun to get used to it here," one resident noted, "so I invite you to come and see my residence." Others wrote: "Here I am in this residential home. Other than that, I am doing quite well"; "I have added a small dot [on the front of the postcard] to mark my floor, room 1.14 … Bye"; and "I am doing reasonably well. I go out for a walk each day and I read a lot." In a humorous tone: "We are doing reasonably well as long as we act like real ladies and don't do too much!" Interestingly, they *did* write about their health, although most of them, in Chowdhury's notes, remained silent on this topic at the dinner table: "We are doing well here … but Herman isn't very fit, it seems like bronchitis"; or "I had a bile attack last week, due to a flu shot" (Van Hamel & Musa, 2019, n.p., all translations KB). Here, the front-stage language that was the norm within the "public" space of the homes was dropped in communication with people who were close to the postcard writers but outsiders to the homes themselves.

The images of the postcards told quite another story. Nearly all photographers used low-angle wide shots (to borrow cinematographic terminology) of the residential homes, positioning their high-rise features center or just off-center. This served to give the high-rise building, the heart of the residential care facilities, pride of place. The linear quality and stillness of the photographs (and this is a more semiotic reading) also flag the architectural modernity and monumentality of the homes. At the same time, the homes hardly ever had a visible environment other than their own garden. Some of the postcards featured images of the homes' interiors, but only a few of these included residents in the frame. Half were men, despite the fact that 75 percent of all residents of these homes were women in 1985 (Chowdhury, 1995, p. 148). Those postcards that did have pictures of interior rooms only showed public spaces such as the dining room, lobby, or billiard room.

These photographs resemble what architectural theorist Sonit Bafna has called "imaginative," rather than "notational," drawings of buildings—depictions that invite an imaginary-perceptual mode of attention to architecture and often create a particular focus, for instance through metaphor (Bafna, 2008, pp. 536–41). Bafna and his colleague Myung Seok Hyun underline that an awareness of the "properties of the medium, of what effects are produced" in the making of architectural photographs helps to read them and to understand their performativity (Hyun & Bafna, 2019, p. 784). Indeed, the postcards that include pictures of the residential home interiors in addition to their exteriors reference postcards of *hotels*, even though *these* hotels were not temporary holiday accommodations but last resorts. Like other depictions of architectural structures, inhabitants are usually absent. Normally, this enables potential residents to project their future selves in the building. In this case, however, the postcards also express something else: by showing units of the same size and blinds of the same color, and presenting the homes without an environment, the pictures underline the anonymity, uniformity, and isolation that the inhabitants attempted to negate through their use of space and postcard narratives (Fig. 1). The postcards

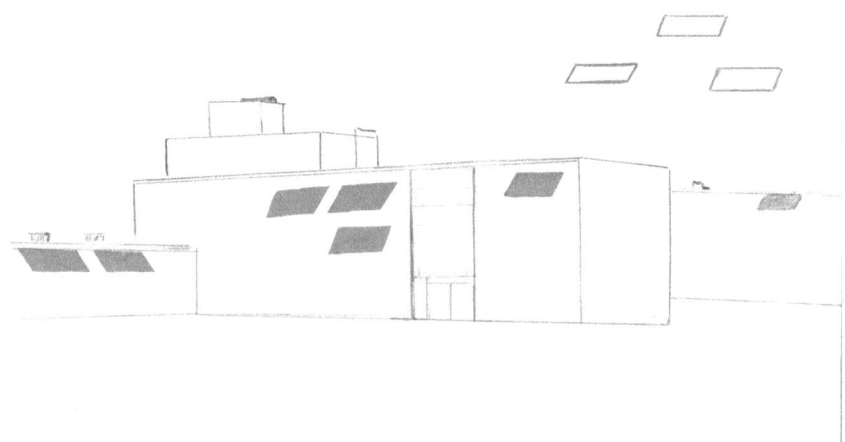

**Fig. 1**  The ritualized space of residential homes © Bijsterveld

thus staged the ritualized space of the residential home as hotel and autonomous suburb but also suggested the type of massiveness that made Chowdhury start from Goffman's work on total institutions in the first place.

## Conclusions

Borrowing insights on spatial rituals from a non-Western anthropologist's study of a Western institution inspired a gestalt-switch in my interpretation of the history of postwar housing for the elderly. Rather than repeating the standard history of a shift in national policies from promoting dependent living to independent living, I was able to show that independent living had already been the ideal since the early postwar years but became associated with changing conceptions of space. Focusing on these spatial ideals, cast in concrete by architects, revealed the ritualized space of residential homes in the concepts and materialities of the single-family home, hotel, village, and suburb.

This anthropological-architectural perspective subsequently helped me to recognize the postcards as a source that could either enrich the existing analysis or potentially provide a fresh point of view. It did both: it confirmed some of Chowdhury's earlier claims about the ways in which residential home residents presented themselves differently to distinct groups of people. The fronts of the postcards, however, highlighted the residential home's hotel-like and suburban features rather than their communal aspects, foregrounding anonymity and isolation. In fact, the postcards may not have functioned as the positive advertisement their makers had in mind, perhaps contributing to the gradual demise of the residential home for the elderly from the mid-1970s onward.

While conducting my research, I did not just develop into an interdisciplinary scholar. Delving into anthropology and, subsequently, sociology and architecture also made me more of a historian. While policymakers stressed discontinuity, I found continuity, a phenomenon historians are, in principle, just as willing to find as discontinuity. I would not have been able to capture the substance of that continuity, however, without familiarizing myself with the anthropological method.

Chowdhury had set the example, though working in the opposite direction: he strengthened his ethnographic understanding of a subculture by incorporating both sociology and the architectural history of housing. This illustrates that interdisciplinarity does not necessarily water down disciplinary virtues and skills—an assumption to which some scholars, as we elaborate in the introduction to this book, seem to adhere when arguing against interdisciplinarity as a scholarly practice. Paradoxically, interdisciplinary work can strengthen such disciplinary virtues and the skills connected to them.

The author of the foreword to the postcard book turns out to be the fictional residential home inhabitant Hendrik Groen. Ironically, he complains that he has not received a postcard for years and that *his* residential home has not issued one. In fact, the end of the television series portrays the closure of the home where Groen had made so many new friends.

# References

Bafna, S. (2008). How architectural drawings work – and what that implies for the role of representation in architecture. *Journal of Architecture, 13*(5), 535–564. https://doi.org/10.1080/13602360802453327

Bijsterveld, K. (1996). *Geen kwestie van leeftijd: Verzorgingsstaat, wetenschap en discussies rond ouderen in Nederland, 1945-1982.* Van Gennep.

Bijsterveld, K., Horstman, K., & Mesman, J. (1992). De versplintering van een categorie: De pensioengerechtigde leeftijd van ongehuwde vrouwen en het rechtskarakter van de verzorgingsstaat. *Beleid en Maatschappij, 19*(1), 16–30.

Bijsterveld, K., Horstman, K., & Mesman, J. (2000). 'Crying whenever Monday comes': Older unmarried women in the Netherlands and the game of comparison, 1955–1980. *Journal of Family History, 25*(2), 221–234.

Bijsterveld, K., Cleophas, E., Krebs, S., & Mom, G. (2014). *Sound and safe: A history of listening behind the wheel.* Oxford University Press.

Chowdhury, S. D. (1990). Privacy, space and the person in a home for the aged. *Etnofoor, 3*(2), 32–47.

Chowdhury, S. D. (1995). Lange levens, kleine kamers. Privacy en afhankelijkheid in een Nederlands verzorgingstehuis. In M. Stavenuiter, K. Bijsterveld, & S. Jansens (Eds.), *Lange levens, stille getuigen: Oudere vrouwen in het verleden* (pp. 148–161). Walburg Pers.

Goffman, E. (1987). *The presentation of self in everyday life*. Penguin Books. (Original Work published 1959).

Goffman, E. (1991). *Asylums: Essays on the social situation of mental patients and other inmates*. Penguin Books. (Original work published 1961).

Greubels, C. (2020). Caring through sound and silence: Technology and the sound of everyday life in homes for the elderly. *Anthropology & Aging, 41*(1), 69–82. https://doi.org/10.5195/aa.2020.229

Groen, H. (2018). *Pogingen iets van het leven te maken: Het geheime dagboek van Hendrik Groen, 83 1/4 jaar*. J.M. Meulenhoff. (Original work published 2014).

Groen, H. (2020). *Opgewekt naar de eindstreep: Het laatste geheime dagboek van Hendrik Groen, 90 jaar*. J.M. Meulenhoff. (Original work published 2018).

Hamel, S., & Muda, R. (2019). *We mogen niet klagen: Kaarten uit het bejaardentehuis*. Uitgeverij De Harmonie.

Hyun, M. S., & Bafna, S. (2019). The photographic expression of architectural character: Lessons from Ezra Stoller's architectural photography. *Journal of Architecture, 24*(6), 778–802. https://doi.org/10.1080/13602365.2019.1684970

Nierstrasz, F. H. J. (1965). *Het grote verzorgingstehuis voor bejaarden in zijn relatie tot het welzijn der bewoners*. Centrale Directie van de Volkshuisvesting en de Bouwnijverheid.

*Nota Bejaardenbeleid*. (1970). Staatsuitgeverij.

*Nota Bejaardenbeleid*. (1975). Staatsuitgeverij.

Oliehoek, T. (Director). (2017–2019). Het geheime dagboek van Hendrik Groen. [TV series]. *Omroep Max*.

Swinnen, A. (2019). Reading ageism in "geezer and grump lit": Responses to The Secret Diary of Hendrik Groen, 83, ¼. *Journal of Aging Studies, 50*. https://doi.org/10.1016/j.jaging.2019.100794

# The Mysterious User of Research Data: Knitting Together Science and Technology Studies with Information and Computer Science

Kathleen Gregory, Paul Groth, Andrea Scharnhorst, and Sally Wyatt

K. Gregory (✉)
Visualization and Data Analysis Research Group, Faculty of Computer Science
& Department of Science and Technology Studies, Faculty of Social Sciences,
University of Vienna, Vienna, Austria
e-mail: kathleen.gregory@univie.ac.at

P. Groth
Informatics Institute, Faculteit der Natuurwetenschappen, Wiskunde en
Informatica, University of Amsterdam, Amsterdam, The Netherlands
e-mail: p.t.groth@uva.nl

A. Scharnhorst
Data Archiving and Networked Services, Royal Netherlands Academy
of Arts and Sciences, The Hague, The Netherlands
e-mail: andrea.scharnhorst@dans.knaw.nl

S. Wyatt
Department of Society Studies, Faculty of Arts & Social Sciences,
Maastricht University, Maastricht, The Netherlands
e-mail: sally.wyatt@maastrichtuniversity.nl

© The Author(s) 2023                                                           **191**
K. Bijsterveld, A. Swinnen (eds.), *Interdisciplinarity in the Scholarly Life Cycle*,
https://doi.org/10.1007/978-3-031-11108-2_11

# Introduction

The open science movement promises to change the production and dissemination of academic knowledge by making the processes and results of research transparent and available, to the benefit of individual researchers, the pocketbooks of funders, and the research enterprise as a whole. Open, accessible, and standardized research data are seen as essential scaffolding for realizing these promises.

Sharing and documenting research data, for example, offer a potential antidote to problems with reproducibility in science by providing a way to validate experimental findings. Reusing research data provides potential economic benefits, by limiting the amount of possibly redundant and nearly always costly data collection. Shared pools of research data offer new possibilities for using data science techniques to tackle society's most wicked problems.

Working to support these visions, data repositories and scientific publishers have increasingly become entangled in mass operations of data documentation and exchange. New tools have been developed to facilitate the discovery of data, and funders and policy makers have implemented policies at national and institutional levels for both open science and data management (European Commission, 2019).

Users are invoked as being central to many of these efforts. Designers of data search tools experiment with sophisticated methods to present the user with the best possible results (e.g., Brickley et al., 2019). Educational tools are designed to help users of repositories and data management tools construct data which are findable, accessible, interoperable, and reusable, or FAIR (Wilkinson et al., 2016). Various metadata schemas, standards, and tools are developed to aid users in discovering and understanding data (e.g., Ohno-Machado et al., 2017).

Despite this stated user focus, the concept of the 'user' or 'users,' similar to that of 'data' and of the practices surrounding data reuse, is conceptualized differently across and within disciplinary domains. In many technical and design-oriented areas of information and computer science, users often remain at arm's length, visible only via ensembles of click behavior, search logs, or data management plans (Van House, 2004).

This acontextual, homogenous view of users contrasts with the heterogeneous, embedded, and socially constructed understanding of use which characterizes work in science and technology studies (STS) (Wyatt, 2003).

Research rooted in these two conflicting views is often undertaken along parallel, yet isolated tracks. When they do intersect, communication between these two perspectives on users is challenging (Tabak, 2014). This chapter reflects upon a project which knit together differing notions of 'users' as a way of grounding interdisciplinary research. In addition to producing novel insights about the reuse of research data, this approach also served to bridge the distance between STS researchers and computer scientists, and between designers of data search systems and users themselves.

After explaining the context of the project, we begin from the end, highlighting the results and outcomes which our 'integrative-synthesis' approach to interdisciplinarity (Barry et al., 2008) afforded. We then turn to the development of our interdisciplinary approach by exploring the conceptual roots of users within information/computer science and STS and discussing how we wove these ideas together within our research.

We conclude by identifying and reflecting on three points that may be applicable to others conducting interdisciplinary research: (i) a common (yet differently conceptualized) idea, for example, 'users,' can serve as an anchor for interdisciplinary work, much in the way of a boundary object; (ii) interdisciplinarity itself is an evolving, contextual construct; and (iii) the broader impacts of interdisciplinary research may change perspectives and practices in ways which are difficult to trace.

## Project Re-SEARCH: Contextual Search for Research Data

Project Re-SEARCH was an interdisciplinary project funded by the Dutch Research Council (NWO, grant number 652.001.002) from 2017 to 2021, which brought together industrial and research partners to investigate and develop search solutions for research data. Researchers from three Dutch universities, a data archive, and an academic publisher pursued three research lines within the project, which were expected to exchange insights and results and eventually resonate with each other. Our STS-infused research line focusing on practices of data discovery and reuse took place alongside research in computer science exploring the development of semantic technologies and relevance ranking algorithms for data search. The academic publisher, Elsevier, provided logistical support for all three lines, with the aim of implementing findings into their prototype search engine for research data, DataSearch.[1]

Each research line consisted of senior and junior researchers performing independent research. The search engine design team at Elsevier varied in composition and size over the course of the project, although on average the team consisted of eight individuals, primarily from computer science. The entire team met monthly to discuss how research findings could be implemented into the data search engine, and how data from Elsevier (e.g., search logs and a dataset index) could be used to inform all research lines.

Researchers working on the project came from a variety of disciplinary backgrounds, including computer science and STS, although typically many more computer scientists and system developers than social scientists or humanities scholars were involved. Even within the broad disciplines of information science, computer science, and STS, the project team had many more specific areas of interest and expertise. In our own research line, for example, the junior researcher had a graduate degree in library and information science, and the senior researchers had backgrounds in computer science, STS, philosophy, economics, and physics.

We brought this multiplicity of disciplinary backgrounds to our research questions and aims which sought (i) to explore how researchers across disciplinary domains discover, make sense of, and reuse data which they do not create themselves, and (ii) to inform and intervene into the development of search solutions for research data.

---

[1] In July 2020, DataSearch was integrated into another Elsevier platform, Mendeley Research Data, available at: https://data.mendeley.com/research-data/

# Beginning from the End: Main Findings of the Project

Understanding what users do—in this case, how researchers locate data for reuse—was a common interest among all team members. Although both information science/computer science (IS/CS) and STS have long histories of exploring how individuals encounter, understand, and engage with information, so-called information-seeking practices, they have done so from different conceptual and methodological standpoints and have only rarely focused specifically on practices related to data.

In the Re-SEARCH project, we embedded STS perspectives about users, communities, and context into established user-centered models of information-seeking common in IS/CS (further discussed in the next section). We knit these two perspectives into an interdisciplinary theoretical construct which we used to frame a range of quantitative and qualitative methodologies, including an analytical literature review, a large-scale survey, observations, and multiple interview studies. Weaving these two perspectives together led to conclusions about data discovery and reuse, briefly outlined in this section, that both spoke to and challenged traditional notions of users and use, particularly within information and computer science.

One of our principal findings centers on the conceptual development of data communities. In the literature on data discovery and reuse, the term 'communities' is often used indiscriminately or to refer to broad disciplinary domains (Borgman, 2012). The results of our research encourage instead a multi-dimensional way of thinking about communities, in which researchers belong to multiple data communities, which are not defined by discipline alone but which rather form around shared data, common data needs, shared methodologies, or common data uses. An example of such a data community can be found in the digital humanities, where researchers from various disciplinary backgrounds come together around a shared (digital) corpus.

Users, of both data and data search systems, are situated within multiple such communities. As they make sense of data for reuse, individuals 'place' data within different contexts, for example, contexts of data creation, disciplinary or social norms, or the data's representativeness of particular phenomena in the world (see Koesten et al., 2021). Our interview study on data-centric sensemaking provides an example of this third type of placing, where study participants worked to place data within the world geographically. Participants questioned if a list of countries in a dataset which we showed them was indeed complete; they also interrogated the granularity of the data, attempting to ascertain if they were representative of an entire country or of only certain areas within a country.

Many of our studies surfaced the importance of data documentation in discovering, placing, and reusing data. Different documentation (e.g., metadata, supporting descriptions, and academic literature) may be needed for different purposes and depend on the 'distances' between users and data in terms of a user's familiarity and expertise. In our interview study focusing on sensemaking, for example, we found that despite their previous knowledge, experienced researchers who are 'close' to the data may need more detailed information than individuals who are 'farther' from the data. This type of detailed information may be best provided using granular, visual representations of patterns rather than the high-level documentation often provided in README files.

We also found that data act as hubs for collaborative activity, as in the case of an early career researcher in the environmental sciences who reported seeking data from other researchers as a way of forming collaborations; data can also provide a means for 'conversations' between data creators and potential reusers. We argue that data discovery systems, repositories, and metadata should be designed to *support* rather than *ignore* the social interactions and collaborative work implicit in discovery and reuse, and we call for innovative solutions, such as interactive forms of data documentation to which both data creators and reusers can contribute.

Finally, our work nuances the idea of 'use,' of both data discovery systems and of data, making visible the multiplicity of actions, resources, and types of data reuse in academic work. We emphasize that both data reuse and the use of discovery systems should be conceptualized as existing on a continuum of uses, rather than as being binary practices of use or non-use. Users of data discovery systems may search for data once or twice online, for example, or they may look for data routinely. Some researchers only reuse data for teaching; others reuse data for multiple purposes, at multiple timepoints in their work. This suggests that although an individual may at times be a user (of a system or of data), they are also, at times, a non-user.

# Knitting Together Differing Notions of Users and Use

The above findings have their theoretical basis in the foundation of our own interdisciplinary approach: the innovative way we knit together the literature on user-centered information-seeking models in information and computer science and the literature on *use, communities,* and *context* in STS. We reviewed these literatures to find points of connection and difference and brought them together into a conceptual framework which guided the rest of our research. To provide insight into this framework and its development, we briefly review these two literatures here and examine how we knit them together as a way of grounding our own interdisciplinarity.[2]

## Information and Computer Science Perspective: User-Centered Models

Cognitive, user-centered perspectives to exploring information discovery have held sway in IS/CS since the mid-1980s (Savolainen, 2007) when research into information behavior was seen to have moved from a systems-oriented view to one foregrounding the information seeker's standpoint (Dervin & Nilan, 1986).

These user-centered approaches to studying information-seeking are rooted in the rather indistinct boundaries between computer science and information science, notably in the fields of information behavior, information retrieval, and interactive information retrieval. Although there are

---

[2] This section draws heavily on the principal output resulting from our research line (Gregory, 2021).

differences between these fields, they tend to converge on their conceptualization of people, who are defined as 'users' of information systems (Jansen & Rieh, 2010).

This view of individuals as users is reflected in numerous conceptual models developed to describe and theorize how people seek information. Such models, many of which date from the final decades of the twentieth century, usually consist of diagrams describing relationships among concepts (Case & Given, 2016), for example, individuals, systems, information, and actions. The majority focus on *search* behaviors (Järvelin & Wilson, 2003), examining how a user interacts with a search system to satisfy an information need.

Information-seeking models have varying levels of specification and serve different research purposes, for example, interpreting observations (Järvelin & Wilson, 2003) or investigating search practices of particular groups (e.g., Ellis & Haugan, 1997). The information journey model (Blandford & Attfield, 2010), which particularly informed our work, synthesizes many key aspects of earlier models. In this model, an information seeker moves through four stages, which are not necessarily required or sequential: *recognizing* a need for information; *acquiring* information, either through active searching, serendipitous discovery, or being told about it; *interpreting and evaluating* information; and *using* the interpreted information.

Across user-centered models, including the information journey model, information discovery is defined through interactions between users and systems but also between contexts and users. Context is a complex and variously defined concept in information research. Positivist views portray context as a backdrop for activities, as an itemized yet inexhaustible list of elements, whereas more relational views see context as being an enacted and local 'carrier of meaning' (Dervin, 1997.) With some exceptions (e.g., Saracevic, 1996; Ingwersen, 1996), established information-seeking models tend toward positivist conceptions, where context is composed of nameable cognitive and affective factors (e.g., goals, tasks, prior knowledge, or feelings such as optimism or uncertainty) which influence a user's information activities (Courtright, 2007). Although many user-centered models start from the view that context shapes search behaviors, they do not address how search behaviors affect broader contexts.

Nancy Van House (2004) summarizes many of the limitations of how users are portrayed in models of information-seeking. Perhaps obviously, user-centered models focus on the viewpoint of one individual acting in a particular role: the person looking for information. This person is defined in terms of interactions with an information system. Other roles that a person enacts, which may influence information practices (IP), are not addressed. Individuals who do not interact with systems are not typically represented. User-centered models do not meaningfully draw out other actors involved in information discovery processes; nor do they account for shared or distributed actions (Talja & Hansen, 2006).

## STS Perspective: Use, Communities, and Context

The division we make between IS/CS and STS might seem a bit artificial. STS-inspired approaches have been taken up within sub-fields of information science, most notably in the area of 'information practices' (IP) research, which examines how information-seeking activities are influenced and shaped by both social and cultural factors (Tuominen et al., 2005), and where information-seeking and use are seen as constructing activities. This area of research reflects the STS tenet that society and technology are not separate entities but are instead co-constituents of a seamless web of dynamic social, material, political, and economic elements (van House, 2004).

In IP research, the focus is not on users per se but rather on the socio-technical infrastructures, practices, and contextual factors surrounding and shaping information-seeking (Savolainen, 2007). The term 'users' is also avoided or invoked with care. For example, social informatics proposes the term 'social actors' to recognize that individuals using technologies are not primarily defined by that use but rather enact multiple roles and inhabit multiple contexts where technologies are present (Lamb & Kling, 2003).

The treatment of users in IP research is mirrored in STS, which has a history of examining the practices of scientists, engineers, and technologists, rather than those of 'users.' When users of a technology are studied, they are viewed as heterogeneous, embedded in dynamic, locally enacted contexts, and are investigated symmetrically. Studies investigate not only how technology shapes user practices but also how users shape the development of technologies (Oudshoorn & Pinch, 2003).

Focusing only on those who engage with a technology ignores the people who do not and thus reinforces the idea that technology use is the norm (Wyatt, 2003). As with users, non-users are not a homogenous group. As Sally Wyatt points out, people may not have access to a technology, they may actively choose not to engage with it, or they may have tried it once or twice and decided that it was not for them. Users should therefore be "conceptualized along a continuum, with degrees and forms of participation that can change" (Wyatt, 2003, p. 77).

Much work in STS also argues that the community or collective, rather than the individual, is the entity that 'knows' (van House, 2004). Knowledge production is situated within these collectives, which have unique norms, practices, and tools. The actions of an individual are representative of these collective ways of knowing and producing knowledge (Knorr Cetina, 1999); the study of individual actors/'users' is therefore inextricably linked with the study of communities. The embedding of individuals and knowledge in communities is also present in information practices research, which emphasizes that individuals act as members of communities and social groups, often in diverse roles, rather than as isolated actors, and that information-seeking is a social practice (Talja & Hansen, 2006).

Work by Karin Knorr Cetina on epistemic cultures (1999) is representative of core ideas about how STS conceptualizes communities. Communities are seen as dynamic groups that are forged through common practices, epistemic norms, and shared objects. Disciplinary domains alone do not define communities. Individuals can belong to multiple communities, and individual actions are shaped by communities themselves. The key to understanding these communities lies in studying practices to reveal actual, rather than imagined, actions and relationships.

## Interdisciplinarity in Practice

In Project Re-SEARCH, we brought the STS perspective described above into user-centered models of information-seeking. A slightly modified version of the information journey model (Blandford & Attfield, 2010), adapted to reflect our research questions, structured our empirical studies. This modified version highlights *users and their needs* and their practices of *discovering, evaluating/sensemaking,* and *(re)using* information, or in the case of our research questions, data. We also drew on tenets of other models (e.g., Saracevic, 1996; Ingwersen, 1996) to understand data discovery as a dynamic process, shaped by a user's purpose or task and involving multiple strategies over time.

Rather than situating data discovery and reuse as isolated practices, we drew on the STS perspectives reviewed above to emphasize the embedding of practice within dynamic networks of people, technologies, materials, and policies. We viewed users not as atomized individuals but rather as social actors enfolded in epistemic communities who engage with technologies in various ways. We worked from the idea that examining objects of research from different perspectives could serve to uncover actual practices of use, to reveal the diversity and multiplicity of practice, and to explore relationships between sociotechnical elements.

To operationalize this perspective, we developed a series of guiding questions (Fig. 1, center). Identifying the various communities, types of data, and technologies used was an important step in understanding data discovery practices. This involved asking 'which' questions, that is, 'which users,' 'which communities,' or 'which data'; it also involved paying attention to which entities were not being taken up or which communities were not engaging in a practice, for example, by identifying absent communities in the current literature or by paying attention to non-response in our survey analysis (Q1). Studying practices of data discovery and reuse required observing what people were doing, as well as questioning how these practices were changing or stabilizing in relation to technologies, materialities, and norms (Q2). Understanding the motivations underlying and the consequences of practice helped to provide context and to trace relationships between particular practices and other elements (Q3, Q4).

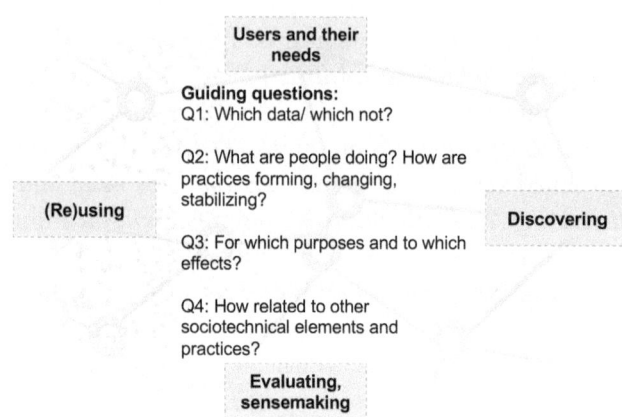

**Fig. 1** Representation of integrated theoretical approach. Boxes represent modified sections of the information journey model from the IS/CS perspective. Guiding questions were developed from key points about use, communities, and context from an STS perspective

Figure 1 visualizes how our conceptual approach was informed by both models and concepts from IS/CS and STS. We further used this interdisciplinary conceptual framework to structure and analyze our empirical work. For example, in our initial study, we conducted an analytical literature review structured along the lines of common features of information-seeking models: user needs, search strategies, and evaluation criteria. At the same time, we analyzed the literature through the lens of a multi-dimensional approach to understanding communities. Rather than focusing on individual data users, we looked for commonalities in practices which might lead to new, emerging groupings of those 'users,' for example, around different types of data.

We conducted two further studies: semi-structured interviews with data seekers and a large-scale global survey. We used the anatomy of the modified information journey model to organize both the interview protocol and the survey questionnaire. In follow-up questions during the interviews and in our analysis of both the quantitative and qualitative survey data, we raised questions about what was present in researchers' data practices. We also sought to identify communities and practices which were not as visible and to tease out relationships between data discovery and other research practices. For example, we asked research participants to explain the similarities, differences, and overlaps between discovering and understanding academic literature versus data, and we encouraged them to explore and untangle the role of professional and personal networks in data discovery and reuse.

## Making Interdisciplinarity Visible

Our interdisciplinary approach is made tangible in the craftwork which we have included at the beginning of this chapter. This data visualization, a 'Knitted Web of Science,' represents the references cited in the principal research output of our research (Gregory, 2021). Wyatt classified all the references, according to her knowledge of the work, and/or by the journal. She then calculated the proportion of the total references for each of the eight disciplinary domains.

These disciplines are knitted in the usual order, starting with the largest, green for information science. Other disciplinary domains in descending order are other natural and life sciences, FAIR and open data, computer science, libraries/archives, methods, STS, and other social sciences and humanities. Our personal areas of expertise are also represented with buttons. The three larger buttons represent the main expertise of Groth, Scharnhorst, and Wyatt. The smaller, pearl buttons spread across the visualization represent the multiple venues where Gregory's work has been published and her command of different disciplinary repertoires.

# Lessons Learned About Interdisciplinary Research

The 'knitting together' of perspectives and methods represented in this craftwork helped us to expose new interdependencies and reach conclusions that would have been difficult to arrive at if we had relied on only one perspective. It also allowed us to apply our findings to develop rec-ommendations for systems design and data documentation and to com-municate those to individuals from various (disciplinary) backgrounds via the language of 'users' and 'uses,' which resonated with a variety of audiences. A perhaps unexpected outcome from this research was the opportunity it provided to reflect on our own understanding of interdis-ciplinarity and to identify points for others to consider when conducting interdisciplinary research.

## A Shared (Yet Differently Conceptualized) Term Is Often an Anchor

IS/CS and STS have approached studying use and users differently, both conceptually and methodologically. It was these differing notions about a common concept that served to ground our interdisciplinary research.

As a concept, 'users' are widely employed within modern software development, where user experience design and testing play a fundamen-tal role (Kashfi et al., 2017). This provided us with a connection point not only with other team members but also with the software developers on DataSearch and with other developers, for example, at the data archive where two of us (Gregory and Scharnhorst) were employed.

The idea of 'users' acting as an anchor point for our work has parallels with Peter Galison's metaphor of the 'trading zone,' where individuals from potentially incompatible viewpoints develop a shared language as a way of bridging differences (Galison, 1997). In our research, we did not match different terminologies or define a shared, working vocabulary, as sometimes happens. Instead, we worked to make explicit the different connotations of a single term. This can be important for interdisciplinary collaboration: if these differences are not made explicit, confusion can arise when people think they are talking about the same concept.

The concepts of 'users' and 'use' acted more as boundary objects in our work (Star & Griesemer, 1989), traveling between different disciplinary and professional communities while being malleable enough to be adapted to fit our research questions and project aims. We then commu-nicated our findings through the frame the 'user' boundary object provided, relying on this concept to reach and resonate with other research and practitioner communities.

## Interdisciplinarity Exists in Different Contexts and Evolves over Time

Not all forms of interdisciplinary research are the same. Interdisciplinarity is enacted through different relationships between contributing disciplines and by various levels of engagement. Andrew Barry and colleagues propose three different modes of interdisciplinarity (Barry et al., 2008). The *integrative-synthesis* mode is characterized by a roughly symmetrical integration of methods and concepts from the involved disciplines. In *service-subordination* mode, one discipline is seen as existing in service to others, contributing without significantly changing the rules of other disciplines (Wyatt, 2021). The *agonistic-antagonistic* mode captures one discipline explicitly aiming to change another. As we saw in project Re-SEARCH, these modes of interdisciplinarity can co-exist and be perceived differently within the same project.

We viewed our research as integrative-synthesis work, bringing together different perspectives from IS/CS and STS in order to enrich the entire project. Our research partners may have had a different view of our role, particularly at the beginning of the project. It could be argued that Elsevier initially saw our STS research line as existing in a service-subordination relationship, where we were expected to adopt the 'correct objective' of the project as a whole, namely to focus on how potential and future users/researchers interact with the DataSearch search engine. At the same time, without Elsevier's engagement, we would not have been able to conduct our own research, particularly the survey, at the same scale or with the same populations.

As the project proceeded, our work challenged the service-subordination model. We did not limit our research to interactions with DataSearch but rather expanded the scope of the problem to focus on data discovery and data reuse practices more broadly, in addition to uses of search technologies. Our STS focus shifted our research line from being object-oriented, focusing on DataSearch, to being practice-oriented, where 'users' provided an entry point to the wider universe of technologies and practices implicated in data discovery and reuse.

Reflecting this, our partners' views of our work evolved as the project progressed. As we deepened our integration of IS/CS and STS perspectives in our own research and as we began to publish our findings and accrued more data, other partners came to see us on a more equal footing. This shift was perhaps accelerated by the fact that we published our work in journals in information science and data science, some of which our project partners had published in themselves. We also collected and analyzed a substantial amount of data in the survey study. These data and our quantitative analysis were perhaps more closely aligned with other partners' own conceptions of what constitutes high quality research. These similarities may have shifted perceptions about our contributions and the role of our research.

## Tracing the Effects of Interdisciplinarity Can Be Challenging

Discussions about tracing the 'impact' of academic work often turn to measures which are easily visible, such as citations. In the short time since we completed our research, we already see signs of these traces. Our studies have been cited in multidisciplinary and discipline-specific journals and in relation to different topics, for example, systems development and data stewardship. Citation practices in different sources, disciplines, and communities vary, which can make it difficult to find and place such citations in the correct contexts.

Another way of viewing the impact of our project could be in the implementation of our recommendations in search solutions for research data, particularly in DataSearch. Changes have been made which align with our recommendations, such as indexing a wider diversity of data repositories and providing links to literature databases. It is difficult to say, however, whether these developments were always planned or whether our research directly contributed to the system's development.

It may take time for the impact of our work to become traceable via such mechanisms. We argue that our work has produced more subtle shifts in practices and perspectives, both in our own team and among others, which are not as visible as impacts documented through citations or new system features.

Signs of these types of impact can be seen in the various invitations for talks and workshop participation which we have received and, also, in discussions around the development of other data search systems, such as Google's Dataset Search. For example, Paul Groth gave a talk on data reuse at the Chan Zuckerberg Initiative; Kathleen Gregory was invited to participate in a Dagstuhl computer science workshop on FAIR data infrastructures. After publishing our survey results, we had conversations with Google Dataset Search about a possible collaboration. Although we did not enter a formal collaboration, a similar discourse to ours can be seen in recent descriptions about Google Dataset Search (e.g., Brickley et al., 2019).

Various team members of Project Re-SEARCH have also reported that their way of viewing data reuse has been altered by our work. Within data archives, especially at the Data Archiving and Networked Services (DANS) where Gregory and Scharnhorst were/are employed, experimenting with different ways of understanding 'users,' particularly as members of data communities, has provided stimulation for moving beyond the idea of organizing archival services only along disciplinary groups.

These less formal signals are indications that our form of interdisciplinarity has helped to bridge the distance between STS researchers and computer scientists, and between designers of data search systems and users. Our approach also has the potential to shape views about 'users' within STS. For example, the models of information-seeking from information science which we drew on, could help to attune STS to different types and temporalities of use. Furthermore, the 'Knitted Web of Science' which we used to introduce and illustrate this chapter (Fig. 2) makes the extent of our interdisciplinary collaboration visible. Knitting one's references may not be a route all researchers could or should adopt, but some awareness of and attention to one's literature and citation practices could help us all to expand our horizons.

**Fig. 2**  The 'Knitted Web of Science' © Gregory et al.

# References

Barry, A., Born, G., & Weszkalnys, G. (2008). Logics of interdisciplinarity. *Economy and Society, 37*(1), 20–49. https://doi.org/10.1080/03085140701760841

Blandford, A., & Attfield, S. (2010). *Interacting with information.* Morgan & Claypool Publishers. https://doi.org/10.2200/S00227ED1V01Y200911HCI006

Brickley, D., Burgess, M., & Noy, N. (2019). Google Dataset Search: Building a search engine for datasets in an open Web ecosystem. *The World Wide Web Conference-WWW '19*, 1365–1375. https://doi.org/10.1145/3308558.3313685

Borgman, C. L. (2012). The conundrum of sharing research data. *Journal of the American Society for Information Science and Technology, 63*(6), 1059–1078. https://doi.org/10.1002/asi.22634

Case, D. O., & Given, L. M. (2016). *Looking for information: A survey of research on information seeking, needs, and behavior* (4th ed.). Emerald Group Publishing.

Courtright, C. (2007). Context in information behavior research. *Annual Review of Information Science and Technology, 41*(1), 273–306. https://doi.org/10.1002/aris.2007.1440410113

Dervin, B. (1997). Given a context by any other name: Methodological tools for taming the unruly beast. In P. Hakkari, R. Savolainen, & B. Dervin (Eds.), *Information seeking in context* (pp. 13–38). Taylor Graham.

Dervin, B., & Nilan, M. (1986). Information needs and uses. *Annual Review of Information Science and Technology, 21*, 3–33.

Ellis, D., & Haugan, M. (1997). Modelling the information seeking patterns of engineers and research scientists in an industrial environment. *Journal of Documentation, 53*(4), 384–403. https://doi.org/10.1108/EUM0000000 007204

European Commission. (2019). *Facts and figures for open research data.* https:// ec.europa.eu/info/research-and-innovation/strategy/goals-research-and-innovation-policy/open-science/open-science-monitor/facts-and-figures-open-research-data_en

Galison, P. (1997). *Image and logic: A material culture of microphysics.* University of Chicago Press.

Gregory, K. (2021). *Findable and reusable?: Data discovery practices in research.* [Doctoral dissertation, Maastricht University]. Maastricht University. https:// doi.org/10.26481/dis.20210302kg

Ingwersen, P. (1996). Cognitive perspectives of information retrieval interaction: Elements of a cognitive IR theory. *Journal of Documentation, 52*(1), 3–50.

Jansen, B. J., & Rieh, S. Y. (2010). The seventeen theoretical constructs of information searching and information retrieval. *Journal of the American Society for Information Science and Technology.* https://doi.org/10.1002/asi.21358

Järvelin, K., & Wilson, T. D. (2003). On conceptual models for information seeking and retrieval research. *Information Research, 9*(1), 1–23.

Kashfi, P., Nilsson, A., & Feldt, R. (2017). Integrating User eXperience practices into software development processes: Implications of the UX characteristics. *PeerJ Computer Science, 3*, e130. https://doi.org/10.7717/peerj-cs.130

Koesten, L., Gregory, K., Groth, P., & Simperl, E. (2021). Talking datasets – Understanding data sensemaking behaviors. *International Journal of Human-Computer Studies, 146*, 102562. https://doi.org/10.1016/j.ijhcs.2020.102562

Knorr Cetina, K. (1999). *Epistemic cultures: How the sciences make knowledge.* Harvard University Press.

Lamb, R., & Kling, R. (2003). Reconceptualizing users as social actors in information systems research. *MIS Quarterly, 27*(2), 197–235.

Ohno-Machado, L., Sansone, S., et al. (2017). Finding useful data across multiple biomedical data repositories using DataMed. *Nature Genetics, 49*(6), 4–7.

Oudshoorn, N., & Pinch, T. (2003). Introduction: How users and non-users matter. In N. Oudshoorn & T. Pinch (Eds.), *How users matter. The co-construction of users and technology* (pp. 1–25). The MIT Press.

Saracevic, T. (1996). Modeling interaction in information retrieval (IR): A review and proposal. *Proceedings of the 59th Annual Meeting of the American Society for Information Science, 33*, 3–9.

Savolainen, R. (2007). Information behavior and information practice: Reviewing the "umbrella concepts" of information-seeking studies. *The Library Quarterly, 77*(2), 109–132. https://doi.org/10.1086/517840

Star, S. L., & Griesemer, J. R. (1989). Institutional ecology, 'translations' and boundary objects: Amateurs and professionals in Berkeley's Museum of Vertebrate Zoology, 1907-39. *Social Studies of Science, 19*(3), 387–420. https://doi.org/10.1177/030631289019003001

Tabak, E. (2014). Jumping between context and users: A difficulty in tracing information practices. *Journal of the Association for Information Science and Technology, 10.*

Talja, S., & Hansen, P. (2006). Information sharing. In A. Spink & C. Cole (Eds.), *New directions in human information behavior* (pp. 113–134). Springer Netherlands. https://doi.org/10.1007/1-4020-3670-1_7

Tuominen, K., Talja, S., & Savolainen, R. (2005). The social constructionist viewpoint on information practices. In K. Fisher, S. Erdelez, & L. McKechnie (Eds.), *Theories of information behavior* (pp. 328–333). Information Today.

Van House, N. (2004). Science and technology studies and information studies. *Annual Review of Information Science and Technology, 38*, 3–86.

Wilkinson, M. D., Dumontier, M., Aalbersberg, I., et al. (2016). The FAIR guiding principles for scientific data management and stewardship. *Scientific Data, 3*, 160018. https://doi.org/10.1038/sdata.2016.18

Wyatt, S. (2003). Non-users also matter: The construction of users and non-users of the Internet. In N. Oudshoorn & T. Pinch (Eds.), *How users matter: The co-construction of users and technology* (pp. 67–79). The MIT Press.

Wyatt, S. (2021). Interdisciplinarity: Models and values for digital humanism. In H. Werthner, E. Prem, E. A. Lee, & C. Ghezzi (Eds.), *Perspectives on digital humanism* (pp. 329–333). Springer.

# Part III

**Cascading Collaborations: With Artists, Style, and Skill**

# Interdisciplinary Anticipations: Art-Science Collaboration at the Maastricht Brain Stimulation and Cognition Laboratory

Flora Lysen

## Introduction

An air of excitement has filled the lecture hall. Dozens of spectators have gathered in the auditorium of the Maastricht University's Department of Cognitive Neuroscience. For some of them, the hall is a familiar professional environment for meeting and interactions with colleagues. Other visitors come from outside the field and have never been in this part of the university or, for that matter, the city. The event marks the end of a first-time collaboration between an artist and the Maastricht brain scientists. Finally, after fifteen months of studying the researchers and their academic practices, the artist will now present her findings.

During that fifteen-month period, some of the people present in the auditorium have grown to like the artist—as a new presence in the research group meetings and lab spaces at the Maastricht Brain Stimulation

F. Lysen (✉)
Department of Society Studies, Faculty of Arts & Social Sciences, Maastricht University, Maastricht, The Netherlands
e-mail: f.lysen@maastrichtuniversity.nl

© The Author(s) 2023
K. Bijsterveld, A. Swinnen (eds.), *Interdisciplinarity in the Scholarly Life Cycle*,
https://doi.org/10.1007/978-3-031-11108-2_12

and Cognition Laboratory. "We really didn't know what to expect," one of the attending scientists told me. She interviewed quite a few researchers inside a glass cubicle in the department's hallway, a former ICT helpdesk office granted to her for the time of her stay. She even volunteered as a test subject and had her brain fitted with electrodes, as well as scanned and magnetically stimulated as part of the lab's ongoing study of visual attention. "A tough cookie," they said. Never asked to exit the scanner. But much of the excited buzz in the auditorium may also be due to participants being puzzled about why their department head chose to spend time on an art-science collaboration. In any case, the general atmosphere is celebratory. And there might be special guests in attendance. Some time ago, members of the university's executive board expressed their interest in the department's collaboration with the artist. They want to know what it could mean for the university. Could this be an "exemplary project" for innovative research?

The 2019 event outlined above marks the end of the first art-science research project funded by the Royal Netherlands Academy of the Arts and Sciences (KNAW) as part of its newly launched art-science platform Mingler. The artist Antye Guenther and neuroscientist Alexander Sack set out to examine alternative ways of conceptualizing and materializing rhythms of brain activity (patterns of electrical pulses generated by neurons) together (Fig. 1 in Chap. 11), while I—having brokered Guenther's and Sack's collaboration from an initial 2018 Mingler match-making event onward—joined incidentally to make sense of it all from my perspective as a cultural historian and science and technology studies (STS) scholar.

In this chapter, I examine the interdisciplinary aspirations of art-science collaborations using the first KNAW Mingler project as an example. From the start, implicit ideas about the potential virtues of interdisciplinarity underpinned Guenther's and Sack's endeavor. In fact, the website of the Mingler art-science platform states that it aims to foster "collaboration beyond the disciplines" and speaks of sharing "knowledge and skills between professionals," as well as sharing "creativity, fascination and dedication" (Mingler, n.d.). Of course, such familiar terms, "creativity," "fascination," and "dedication"—and we may add "innovation," "collaboration," and "co-creation"—are virtue words (also called "ideographs," cf. Van Lente, 2000), bound up with the indefinite norms and values about research that shape notions of "interdisciplinarity," a concept that is especially prevalent in the art-science domain. Studying the unfolding of an art-science project up close allows a view of the way such ill-defined commitments develop, and how what we may call "imaginaries of the 'inter'" are shaped, that is, how such imaginaries are, to paraphrase Sheila Jasanoff, collectively held, institutionally (un) stabilized, and publicly performed (2015, p. 4). Such a situated approach may help to counteract an enthusiastic but frustratingly vague "normative weight," carried by the prefix "inter," as sociologist Felicity Callard and Des Fitzgerald explain in *Rethinking Interdisciplinarity Across the Social Sciences and Neurosciences* (2015, p. 4). As argued by the authors, eagerness about interdisciplinarity may eclipse a critical examination of the conceptual assumptions, institutional mechanisms, and concrete actions that inform calls to cross disciplinary boundaries. In their words, "interdisciplinarity is a term that everyone invokes, and no one understands" (p. 4).

In this chapter I take heed of such critiques of ubiquitous, yet nebulous calls for interdisciplinarity, and I aim to capture some of the social, epistemic, and symbolic operations of this slippery term in action. Tracing the Maastricht Mingler art-science collaboration from its start to its (provisional) end, I am particularly interested how interdisciplinarity impacts "epistemic living spaces," as Ulrike Felt calls the multi-dimensional structures that shape how research is (and can be) done and how one can be a researcher (Felt, 2009). Based on the notion of epistemic living spaces, it is possible not only to pay attention to policy discourses and social imaginaries that influence epistemic cultures but also to call for attention to other, tacit structures, such as the more "implicit dimensions of 'being in a field'" and the subtleties of interpersonal relations (p. 20).

As a participant-observer of Guenther's and Sack's project, my own expectations inevitably pervade my analysis. Therefore, I start this chapter by contextualizing my initial enthusiasm for collaboration in the field of art-neuroscience. I first describe how such collaboration can be seen as part of a boom in art-science projects, while also representing gleeful hopes for potential insights to be generated from interdisciplinary research between the neurosciences on the one hand and the social sciences and humanities on the other. Secondly, I reflect on the process of match-making through Mingler and my presumptions in brokering an art-science collaboration between an artist and a neuroscientist. Moving on to describe the shifting status of "collaboration" in the project, in the third section of my chapter, I point to the pervasive presence of anticipation and confusion as typical affects in interdisciplinary work. Finally, describing the grand finale of the art-science project, I note a shifting of established hierarchies that can momentarily take place within the space of a collaboration. During Guenther's final performance at the faculty, an underdetermined and fuzzy set of expectations or "potencies" around joining different disciplines allowed for a playful destabilization, as implicitly invoked by her title: "MAASTRICHT TRANSFORMATIONAL SUPERNODE GATHERING OF INTELLIGENT MINDS: No Body, Never Mind — How to Beautify Your Brain Data and Use it to Unleash Your Full Potential."

## Interdisciplinary Aspirations, High Hopes

My initial personal expectation about the interdisciplinary potential of a new art-neuroscience collaboration was high. In the past fifteen years or so, I had observed and studied many interesting art-science projects as part of a surge of research-oriented efforts at the intersections of art and the brain sciences. From cognitive scientists collaborating with dancers to conceptualize synchronicity in brains and interacting bodies (Mutual Wave Machine, Suzanne Dikker, and Matthias Oostrik), and film makers working with synesthesia researchers to emulate the feeling of sharing a sensation with an object (Sensorium Tests, Daria Martin), to STS researchers working with cognitive scientists to create an "experiment-performance" that questions the established protocols of a psychology experiment (Klein & Margethis, 2017). Each art-(brain)science project assembles a very particular set of disciplinary expectations and institutional architectures.

As pointed out by sociologists Andrew Barry and Georgina Born, art-science can function as an exemplary field to study the dynamics and politics of interdisciplinarity (Barry & Born, 2014). Drawing on their work, I have studied several art-science projects to examine how they allow different disciplinary relations—critical, explorative, celebratory, ambiguous, even though such adjectives are clumsy and imprecise—among the arts and vis-à-vis the brain sciences (Lysen, 2019). Art may take on the role of ethical or critical commentary on research, for example, on the enduring hype surrounding neuroscientific research. It may offer an approach to social engagement in academia, a way of fostering the marketability of a science or a technology, a form to address affective and ineffable elements in research, a method for science to become more methodologically reflective—as well as all of the above, to different degrees, at the same time.

Analyzing such intersecting dynamics in art-science projects, Born and Barry provide a valuable conceptual taxonomy of three main "logics of interdisciplinarity": a "logic of accountability" through which art-science works are meant to stimulate debate about and engagement with science, a "logic of innovation" through which art-science is situated as a partner in providing new insights for innovation (two logics that largely render art subservient to science), and a "logic of ontology" that may constitute a true hybridization of fields (Born & Barry, 2010). The ontological dimension is present when art-science practices redefine the object of research and the subjects and publics engaging with it, contributing "to the generation of something new within scientific practice itself, challenging the boundaries of disciplinary authority" (p. 114). For this reason, accountability, innovation, and a type of generative disciplinary disruption feature as central expectations attached to interdisciplinary in art-science projects.

Today, interdisciplinary forms of art research are booming in particular. The field of art-science, it seems, has finally moved away from its long-time "nascent" status to become a more mainstream phenomenon. "Scientists and artists are working together as never before," the journal *Nature* headlined in 2021, dedicating a number of articles to the phenomenon (The Editors of Nature). Since the 1990s, art-science residencies and art-technology collaborations have become increasingly institutionalized and professionalized (Wisnioski & Zacharias, 2014). By now, it is impossible to list the number and variety of collaborative platforms, residencies, funds, foundations, and institutions that allow interfaces between arts and research. This is also evidenced in the expanded vocabulary used to refer to art-science collaborations: "sciart," "artsci," "bioart," STEAM (science, technology, engineering, arts, mathematics), SEAD (science, engineering, arts, design), art-science, art & technology, artistic research, research creation—not to mention subfields such as art and medicine and the medical humanities. All of these terms come with different conceptual inflections and (institutional) histories of course.

This context explains my specific interest in participating in an art-science project, which was triggered also by a more recent addition to the heterogeneous set of art-science infrastructures and projects: the field of "Art and Science and Technology Studies," abbreviated "ASTS" (Rogers & Halpern, 2021, cf. Borgdorff et al., 2019). By adding the term "Arts" to the existing discipline "STS," a discipline which itself emerged out of interdisciplinary activities, ASTS rhetorically positions itself as a new discipline, even though Hannah S. Rogers and Megan Halpern prefer to speak of a "framework," "an emerging way of knowing," or a "new knowledge field" that examines art-science across the natural or life sciences, the social sciences, the humanities, and the arts, using STS as a methodological lens but adding artistic methods to STS at the same time (Rogers & Halpern, 2021, n.p.). The "A" in ASTS, then, denotes both an object of study (projects that interface art and other disciplines) and a methodological innovation. Artists, in Rogers's view, may be making "STS arguments" by "material means" and in tandem, while established STS methods can be enriched by research in and through the arts. In fact, a number of authors in the *Handbook of Science and Technology Studies* (Felt et al., 2016) argue that if STS was more open to research through the arts, STS itself would become more experimental; it would not only observe people "thinking with eyes and hands," but "[use] eyes and hands to intervene and interfere in spaces and sites where science and technology are constructed, distributed, used, incorporated, and enacted" (Salter et al., 2016, p. 154).

So, it is this recent attention to ASTS, combined with a general boom in art-science work, that had my interest, which met with another, parallel development: the call for more interdisciplinary research into the human brain. Indeed, while dreams of new synergies and cooperation loom large everywhere in academia, the pervasive promise of inter- or transdisciplinarity has perhaps been important in particular for the field of neuroscience (Callard & Fitzgerald, 2015)—a field typified by some as a "hybrid of hybrids" (Abi-Rached & Rose, 2010). Scholars subscribing to an emerging field of Critical Neuroscience, for example, have cautiously suggested that some forms of inter- or transdisciplinarity may be a way forward in conducting richer, more nuanced neuroscientific experiments with an "ethos of reflexivity," designing and conducting brain research that is aware, for example, of the complex interchanges between biological and social developments (Slaby & Choudhury, 2017). In fact, it is suggested that interdisciplinarity research by humanists and social scientists together with scientists could perhaps be a means to arrive at "a more expansive account of human development and subject formation" (Frost, 2018), as well as to counteract a reductionist understanding of the brain and human behavior.

At the same time, such high hopes for interdisciplinarity are also met with skepticism, as they may leave unacknowledged the power asymmetries between the authoritative and well-funded discipline of the neurosciences versus the publicly less-prestigious and underfunded disciplines of the humanities, social sciences, and the arts, which are thus prone to be cast in a subservient role. Moreover, there are few accounts of what interdisciplinarity actually *does*. While there are many calls for interdisciplinary research, the actual procedures and effects of engaging novel disciplinary relations are often left underexamined (Fitsch et al., 2021). In this respect, Callard and colleagues emphasize the importance of studying closely the actual configurations of multidisciplinary practices to adjust ideal-type descriptions of collaboration and boundary work and to gain more insight into the unfolding of "science-and-humanities-and-arts-in-the-making" (Callard et al., 2015, p. 4). Thus, it was with a call to study "science-and-humanities-and-arts-in-the-making," as well as with an eye to ASTS and the curbed incredulity of interdisciplinarity, that I set out to participate in one of the first match-making events for artists and scientists in the Netherlands, the initial KNAW Mingler evening.

## Match-Making, Co-laboration

I met the scientist in a room with stucco ceilings and gilt-framed mirrors in the seventeenth-century building of the Royal Netherlands Academy of the Arts and Sciences (KNAW), right in the center of Amsterdam. He was my top choice out of a number of potential matches during an evening organized as part of the Academy's new art-science Mingler collaboration platform encouraging "interaction and synergies in scientific and artistic research" (Mingler, n.d.). For the time being, the platform was open by invitation only, a privilege new members could extend to three new invitees. In Mingler's online interface, participants are prompted to describe general interests and to select (from a standardized drop-down menu) academic and artistic disciplines to be matched with. In a tongue-in-cheek fashion, a visual interface also allows participants to select the characteristics of a fitting collaborator, choosing between affinities for "thinking" and "doing," "details" and "bigger picture," "process" and "result," "risk" and "certainty." Reducing these categorizations to simple buttons on a mix panel somewhat ironically hinted at the impossibility of quantifying the process of (inter) disciplinary "mingling."

My match, professor Alexander Sack, head of the Brain Stimulation and Cognition laboratory and research group at Maastricht University and an expert in the field of transcranial magnetic stimulation, and I sat on plush chairs to discover our mutual interests. We talked about the portrayal of neuroscientific research in popular media and the allure of colorful brain visualizations that journalists and neuroscientists—we both agreed—used to amplify the power of brain-centered explanations of social and cultural phenomena. With witty irony, Sack lamented the fact that such pretty pictures were sadly lacking in most of the brain rhythm measurements he carried out in his lab. His group's focus was on using non-invasive brain stimulation to understand basic mechanisms of perception and attention, as well as on research into the clinical application of brain stimulation to treat patients with severe depression. Clearly, even without attractive brain images, brain stimulation was a mesmerizing topic. Sack showed me a video of a black magnet hovering just above the head of a person speaking, who suddenly stopped mid-sentence when the magnet was activated: "Humpty-dumpty sat on a wall, humpty-dumpty had a great fff[…]" (Read, 2011). The field of brain stimulation research conjured its own captivating science-fictional imaginaries that invoked speculations of mind control and anxieties about future misuse of this technology (Rose & Rose, 2016). I told professor Sack about artists but also social scientists and humanities scholars (like myself) who would perhaps be interested to study the way the lab workers engaged with and perhaps even participated in such imaginaries. Our match-making had worked: we agreed to continue our conversation and to think about a collaborative project, perhaps inviting an artist to work with us.

Only later, after more interactions with the scientists in Sack's research group, I realized the presumptuousness of my initial proposal to "bring" critical artistic and STS insights to the lab. Anthropologist Jörg Niewöhner, analyzing anthropological research into natural sciences, has characterized this attitude as a mode of "critical engagement": a way of relating that predominantly aims to deconstruct "the epistemic regimes to reveal illegitimate reductions of the richness of human group life to material quantities" (2016, p. 1). He recognizes this attitude in particular in STS projects, often geared toward deconstructive critique, even though STS scholars "by turning their revelation of contingency into propositions for the field" may "hope to produce a productive intervention" (p. 16). However, in the months to come, I would begin to see how the Maastricht lab members were not the reductionist researchers in need of extra-disciplinary insights and productive interventions that I too had somehow imagined them to be.

For one, the Maastricht Brain Stimulation and Cognition laboratory had a longer history of hosting researchers from other disciplines. Not long ago, a philosopher of technology had videotaped interactions with the brain-stimulating magnetic device and had sat in on numerous lab meetings. In addition, three lab researchers had taken the issue of the dystopian visions attached to neuro-enhancement head-on in a scientific article, proposing alternative ways to conceptualize the ethical threats posed by fundamental stimulation research (Duecker et al., 2014). Lab members self-organized reading groups in philosophy of science and neuroscience, while discussions about epistemological issues in brain-imaging—throughout all our conversations—were quick to surface. Certainly, these scholars developed critical perspectives from within the field—without extra-disciplinary visitors needing to contribute smart, snappy commentary. Therefore, realizing the self-reflexive attitudes and practices in the lab, I experienced the classic anxieties of an anthropologist going native, if not the sense captured by Niewöhner when asking whether "the actors in the field knew all along what the anthropologists proudly present to them as their findings" (2016, p. 3).

To reconceptualize this issue of expert anthropology, Niewöhner proposes to cease thinking of the anthropologist as possessing some special kind of reflexivity. Rather, the anthropologist can work to strengthen the spaces and infrastructures that allow "reflexing" (practicing reflexivity) by all actors involved. Niewöhner proposes the term "co-laboration" for this model of "joint epistemic work, experimenting with formats without necessarily aiming for a shared goal" (2016, p. 10). Specifically in the context of art-science, Niewöhner's concept of co-laboration as a space for joint—but not exactly united—investigations may offer a significant alternative to an imagined collaborative vision of interdisciplinarity, of shared work, between artists and scientists.

But how can such joint co-laboration be facilitated? Callard and Fitzgerald note how a "rhetorics of reciprocity and mutuality" pervades the literature on interdisciplinarity and shapes an image of interdisciplinarity as collaboration based on "fair exchange" or a "fantasy of equal actors" (2015, p. 100). In practice, such mundane realities as funding rules are important determinants for the organization of art-science collaborations (Boehm, 2018): who visits who? What counts as a final result? Who determines the vocabulary for communicating about the event? And, perhaps, how do artists need to frame their work *as research* to be considered eligible for funding? For example, when in 2018 the KNAW first announced its first Mingler grant as an incentive for "starting art-science collaborations," it called for the roles of the artist and the scientist to be "balanced," but elegantly left open the exact nature of the collaboration in the grant applications rules, which asked to describe "the way in which different needs, perspectives and methods of the arts and science come together" (Regulations for Mingler scholarship, n.d.). Nevertheless, only artists were entitled to a personal allowance paid from the grant.

It was after considering these conditions that I decided to broker between Sack and the artist Guenther, known to me for her critical work on the circulation of scientific images and imaginaries of brain control and for her creation of ceramic objects that give shape to abstract concepts as part of narrative installations. After introducing the two and some emailing back and forth, they did actually find mutual ground to apply for the Mingler grant together. Guenther proposed to investigate scientific practices and ideas at the brain stimulation laboratory for a fifteen-month research period. Together with Sack, she would investigate how researchers in the lab envisioned complex patterns of neural activity in the human brain, which would lead to a "speculative manual" proposing new ways of conceptualizing, visualizing, and/or materializing these brain rhythms (Fig. 1). My role as a cultural and social scientist was to engage in a second-order observation of Guenther's and Sack's intended art-science collaboration.

From the outset, it seemed evident that an interesting art-science project would need to steer away from the tendency to assign the artist the instrumental role of *visualizing* science "post-closure." Instead, one of the initial questions Guenther's and Sack's art-science project set out to examine was how scientists working in the Maastricht lab may be creating some implicit working concept of a complex pattern in order to engage in researching different aspects of brain stimulation. What images, phrases, metaphors, gestures, and materials were drawn upon to work with this oscillatory "unknown" at the center of brain stimulation research?

Sack, reflecting on his motivations for starting the art-science collaboration during a symposium on interdisciplinary research in the brain sciences, described his wish to participate as a way to transcend familiar epistemic and conceptual cognitive science conundrums (the much-discussed problem of "mental representations," for example) and move away from established habits and jargon in his lab: "I had the feeling it might be good to step out of this bubble and to talk to someone with a different perspective (…) with a completely different way of relating things,

**Fig. 1** Research collage by Antye Guenther, based on MRI brain data visualizations assembled at the Maastricht Department of Cognitive Neuroscience, 2022 © Guenther

someone from a different field" (Fieldnotes Lysen, Brain Culture Interfaces workshop, 2019). What surprised Sack was Guenther's approach of investigating, "rather than a passive person in the lab looking at us from the outside trying to judge, challenge and to help … you wanted to become a part of the group." For Guenther too, this was a new way of working, to apply together for a grant, to draw up a plan together, we "somehow made an unwritten contract … we are both active." Key to her work in the project was to go behind the public image of the lab, beyond publications and lectures. But this different way of engaging, according to Guenther, also came "bittersweet": no longer could she just "gather some things and go." The intensity of her presence meant high expectations were building up on the part of the scientists.

## Anticipatory Feelings, Ambiguous Affects

Keeping track of this art-science project and meeting with Guenther and Sack from time to time, I witnessed the increasing embeddedness of the artist in the neuroscientific research environment. Guenther regularly participated in the group meetings, and she clearly knew her way around. She was acutely aware which projects the various researchers were working on specifically, what rooms people occupied, which deadlines they were trying to meet. The official KNAW-funded status of the project helped her to feel more comfortable being present in the lab, she said, allowing for a helpful sense of entitlement. Working together directly with Sack, the head of the department, also added to this sense of legitimate presence. Another important aid in levelling the playing field for her presence in the lab was the artist's previous expertise as a medical doctor—she had the outsider status of an artist paired with the credentials and background knowledge of a field much closer to the cognitive sciences.

Pursuing the project's initial research question—how do lab researchers imagine patterns in (the oscillations of) brain activity?—Guenther noted how scientists created gestures (wavy hand motions), used graphic notations (frequency bands), or employed metaphors (orchestra's) to make sense of the basic neural mechanisms under investigation. And, yes, these images were of course restrictive, one of her interlocutors agreed, "it's difficult to operationalize the questions one has … you have to simplify things to isolate the things you want to see. … What we see depends on what we already know" (Fieldnotes Guenther, cited during Brain Culture Interfaces workshop, 2019). Conducting interviews with lab members—from students to PhDs, postdocs, and senior researchers—Guenther was surprised, she later relayed, about the wide range of opinions and reflections on the state of the field, even in this very tight-knit and collaborative environment.

Gradually, her investigation spread into many different directions. Like a magpie in the lab, a self-proclaimed "scavenge hunter," Guenther's method was to notice, to note, and to accumulate. Every now and then, she would share her finds: photographs of DIY lab set-ups, quotes from the neuroscientists she had interviewed, YouTube videos the lab researchers had used to explain a scientific phenomenon to her, materials and images from an adjacent fMRI-research unit, and goodies from a European Human Brain project conference. But when I myself noted a certain proximity to methods in laboratory ethnography and STS approaches, Guenther refused those labels, emphasizing instead her ultimate goal to produce art, her responsibility, however uneasy, for the final result of the project to take the form of a "work," as she put it. And in fact, during the process of assembling images and impressions, Guenther's presence was building up aspiration and expectations in the neuroscientists, too. Everyone in and around the research unit was aware the project was to culminate in some artistic format. Reflecting on her ambiguous position in Sack's lab, Guenther noted: "I feel they are all contributing to my practice. I hope ... that I'll meet some of the expectations. That's my worry that I'm only taking. ... I'm grateful for the time and commitment they give me. But I'm afraid to ask them what they get out of it" (Fieldnotes Lysen, Brain Culture Interfaces workshop, 2019).

Guenther's worries about reciprocity and anticipations are characteristic of the ambivalent feelings at play in (envisioned) interdisciplinary spaces. Callard and Fitzgerald (2015) emphasize that paying attention to such affective dispositions is key to understanding temporary social spaces of collaboration: unspoken distrust, power unbalances, productive vagueness, and a sense of awkwardness and ignorance, for example. In their own interdisciplinary experiments, the authors most often observed what they call "feeling fuzzy," a "feeling of confusion about what one is feeling" in the practice of working together (p. 115). In the context of the Maastricht art-science experiment, this consideration of the ambiguous affects of interdisciplinarity helps to better understand how uncertainties about the process and goals of the endeavor could be accommodated by the framework and the process of collaboration.

Puzzled feelings about the nature of the exchange ("what they get out of it") demonstrate the strange inversion of hierarchies that can take place within the space of an art-science project. Barry and Born (2014) describe how some actors may envision art-science projects as providing a service to science, an instrument in making opaque and complex processes more approachable for a lay audience (the aforementioned "logics of accountability"). Yet, within the microsocial space of some art-science collaborations, these roles can also be partially inverted (Born and Barry give an example from the 1980s, when the researchers of the French Institut de Recherche et de Coordination Acoustique/Musique [IRCAM] directed much of their scientific and technical force into the preparation of Pierre Boulez's music piece *Repons)*—instead of artists serving science, scientists can offer resources for autonomous artistic projects (p. 12).

At least for Sack, so it appeared, it was the artist's continuous active presence as embedded outsider and the process of attuning to that presence by the lab members that constituted a major part of the perceived value of this art-science project. This meant the Maastricht Mingler project was characterized by a peculiar disjunction. On the one hand, the project entailed the "open," "shared," and "inquisitive" process of an artist aligning with the collaborative style of working in a laboratory research group. But on the other hand, Guenther's simultaneous solo practice, being equally central to the project, was situated alongside the laboratory (in the artist's laptop, in her studio, and in her mind)—a practice relatively opaque and closed to the researchers—which secured the autonomy of the artist in this art-science alliance. Ultimately, it seemed that the enduring uncertainty regarding the project's final outcome did not bother the lab workers so much as it added to a welcome sense of positive excitement. A date for a final presentation had been set in the research group's calendar. They were looking forward to "it."

## Grand Finales, Exceptional Powers

Expectations had run high indeed. Although it felt as if the project had only just started, the academic funding scheme specified that the Maastricht art-science project needed to end. Guenther picked the format of a performative lecture as a fitting medium to assemble the array of narratives, images, and objects gathered during her fieldwork. Sparked by her finds and observations at the Maastricht lab, she created white porcelain brain-shaped vases based on 3-D images. These delicate-looking oddities functioned as "props" for her final performance entitled, as said, "MAASTRICHT     TRANSFORMATIONAL     SUPERNODE GATHERING OF INTELLIGENT MINDS No Body, Never Mind — How to Beautify Your Brain Data and Use it to Unleash Your Full Potential," which was staged multiple times in October 2019. By choosing the genre of performance, Guenther could appropriate and subvert the postures, movements, and explanations she had encountered in the lab and wider sphere of neuroscience, playfully alluding to the wealth of popular science lectures and TEDtalks featuring brain scientists.

Situating the performance in the auditorium of the Maastricht neuroscience department, Guenther subtly transformed this academic space into a stage: adding theatrical stage lights to the existing technical infrastructure and wearing a custom-made dress from exactly the same sound-proof material as the backwall of the room. While the Maastricht neuroscientists had mostly warded-off associations with brain stimulation as a form of cognitive enhancement—wary of science-fictional exaggeration and hype—Guenther reintroduced those associations, bringing para-scientific worlds back into the space of the department. Throughout the performance, Guenther shifted between characters, performing fragments of commentaries that to me seemed hints of a scientist performing an experiment and of an archaeologist of the future, excavating remainders of society that had suffered total data annihilation, trying to make sense of an artist's notebook found in a Maastricht University department.

Observations on experimenting and experimentation continuously intersected in the performance. Guenther made astute and witty comments on her own experience of lying inside a scanner, much in line with the work of STS researchers questioning neuroscientific research paradigms while participating in a brain recording (Roepstorff, 2001; Langlitz, 2013). Moreover, the spectators present in the auditorium were themselves cast into the role of experimental subjects ("let's synchronize our brain waves"), subjugated to a subtle protocol of subliminal influencing, hypnosis, and priming in which the artist took the role of the authoritative scientist-motivator puppeteer, clearly at the top of the disciplinary food chain. Now, for this one hour of performance, Guenther ironically addressed the power dynamics of art-science, at one point framing the event as a match-making where neuroscientists could find "highly motivated" visual artists, who

> do not only come with evidently sharp minds and these exceptional powers of imagination … No one takes them really seriously. Everyone enjoys their exotic presence, they spark everyday routines, while opposing no real threat to no one. And this perception is crucial to us, as it opens so many otherwise closed doors. (Performance notes by Antye Guenther)

Direct commentary on the project's uneven foundation was paired with accounts of scientists trying to visualize concepts and jokes about the hubris and omnipresence of the brain in pop culture.

I experienced the performance as a mesmerizing puzzle movie, a complex narrative of clever analytical pieces, in which I sensed a pattern but could not grasp it—just yet. In the days and months after the project's grand finale, one neuroscientist-spectator told me he saw the piece twice to try to get a better understanding of its structure and dialogue. Another visitor from a different department lamented not having been part of the process, to "see all the connections," and hopes the project will find a second iteration at his workplace. Some of the reactions to this artistic finale—a feeling of bewilderment, of not knowing, but excitement over its collaborative audacity—reverberate again the ambiguous affects that characterized the process of this art-science project all along. Ultimately, what remained after the performance was a sense of potency—a new mode of working had (only just) begun to emerge and needed further exploration.[1]

---

[1] And, indeed, collaborations continued after the Mingler art-science project in Maastricht. Sack, Guenther, and I wrote a visual-textual exercise in "interdisciplinarity" together, which we hope can be helpful for other collaborators in art-science (Guenther et al., forthcoming). Guenther built on her work at the neuroscience department to start a new research project as the first PhD candidate in artistic research affiliated to Maastricht University as part of the MERIAN research in the arts network.

## Conclusion

While notions of collaboration and co-creation often cast interdisciplinarity in a romantic light and suggest a productive and innovative "symbiosis" of disciplines is possible, my above-discussed analysis of a specific trajectory of art-science collaboration shows that uncomfortable affective dispositions may result from a temporary joint project. As a participant-observer of Guenther's and Sack's attempt at art-neuroscience collaboration, I have noted momentary inversions of hierarchies and a-synchronicities: the scientists at the Maastricht Brain Stimulation and Cognition laboratory who facilitate an artist who is trying hard to live up to rising expectations; the artist who carves out a space for autonomous artistic practice parallel to group participation; and the potential of the performance medium to allow—if only for a very brief moment—a switching of the established balance of power.

My analysis revealed how the open-ended structure of Guenther's and Sack's art-science collaboration allowed for a shift in focus not so much on a material effort of co-creation, but on the presence of the artist in the research spaces of the neuroscientists. A growing feeling of anticipation for a "final artwork" was an important part of this trajectory. It was precisely this affective structure of anticipation that opened up prospects of resistance and allowed participants to play with hierarchies in unexpected ways. The tacit affective dispositions I describe in this chapter demonstrate the implications (ideals) of interdisciplinarity in particular situations, beyond a mere discursive analysis of imaginaries of the "inter." Rather than ask "how is this art-science project interdisciplinary?" I have traced a process of "science-and-humanities-and-arts-in-the-making": The first KNAW Mingler art-science project in Maastricht that I have analyzed here underscores how different actors—including myself—shape aspirations for—and anticipations of—*doing* interdisciplinarity.

# References

Abi-Rached, J. M., & Rose, N. (2010). The birth of the neuromolecular gaze. *History of the Human Sciences, 23*(1), 11–36.

Barry, A., & Born, G. (2014). Interdisciplinarity. Reconfigurations of the social and natural sciences. In A. Barry & G. Born (Eds.), *Interdisciplinarity: Reconfigurations of the social and natural sciences* (pp. 1–56). London: Routledge.

Böhm, B. (2018). From heterogeneity to hybridity?: Working and living in arts-based research? In P. Sormani, G. Carbone, & P. Gisler (Eds.), *Practicing art/science: Experiments in an emerging field* (pp. 125–141). Routledge.

Born, G., & Barry, A. (2010). Art-Science: From public understanding to public experiment. *Journal of Cultural Economy, 3*(1), 103–119. https://doi.org/10.1080/17530351003617610

Callard, F., & Fitzgerald, D. (2015). *Rethinking interdisciplinarity across the social sciences and neurosciences*. Palgrave Macmillan.

Callard, F., Fitzgerald, D., & Woods, A. (2015). Interdisciplinary collaboration in action: Tracking the signal, tracing the noise. *Palgrave Communications, 1*. https://doi.org/10.1057/palcomms.2015.19

Duecker, F., de Graaf, T. A., & Sack, A. T. (2014). Thinking caps for everyone? The role of neuro-enhancement by non-invasive brain stimulation in neuroscience and beyond. *Frontiers in Systems Neuroscience, 8*. https://doi.org/10.3389/fnsys.2014.00071

Felt, U. (2009). Knowing and living in academic research. In U. Felt (Ed.), *Knowing and living in academic research: Convergence and heterogeneities in European research cultures* (pp. 17–39). Institute of Sociology of the Academy of Sciences of the Czech Republic.

Felt, U., Fouché, R., Miller, C. A., & Smith-Doerr, L. (Eds.). (2016). *The handbook of science and technology studies*. MIT Press.

Fitsch, H., Lysen, F., & Choudhury, S. (2021). Editorial: Challenges of interdisciplinary research in the field of critical (sex/gender) neuroscience. *Frontiers of Sociology.*. (forthcoming).

Frost, S. (2018). Ten Theses on the Subject of Biology and Politics: Conceptual, Methodological, and Biopolitical Considerations. In M. Meloni, J. Cromby, D. Fitzgerald & S. Lloyd (Eds.), *The Palgrave Handbook of Biology and Society* (pp. 897–923). Basingstoke, Hampshire: Palgrave Macmillan.

Guenther, A., Lysen, F., & Sack, A. (forthcoming). Circulating neuro-imagery – an interdisciplinary exercise. In S. Besser & F. Lysen (Eds.), *Worlding the brain. Interdisciplinary explorations in cognition and neuroculture*. Brill.

Jasanoff, S. (2015). Future imperfect: Science, technology, and the imaginations of modernity. In S. Jasanoff & S.-H. Kim (Eds.), *Dreamscapes of modernity: Sociotechnical imaginaries and the fabrication of power*. University of Chicago Press.

Klein, S. A., & Marghetis, T. (2017). Shaping experiment from the inside out: Performance-collaboration in the cognitive science lab. *Performance Matters, 3*(2), 16–40.

Langlitz, N. (2013). *Neuropsychedelia: The revival of hallucinogen research since the decade of the brain*. University of California Press.

Lysen, F. (2019). Kissing and staring in times of neuromania: The social brain in art-science experiments. In T. Pinch, H. Borgdorff, & P. Peters (Eds.), *Dialogues between artistic research and science & technology studies* (pp. 167–183). Routledge.

Mingler. (n.d.). Retrieved September 16, 2021, from https://mingler.network/

Niewöhner, J. (2016). Co-laborative anthropology: Crafting reflexivities experimentally. In J. Jouhki & T. Steel (Eds.), *Etnologinen tulkinta ja analyysi: Kohti avoimempaa tutkimusprosessia* (pp. 81–124). Ethnos. Reprint in English translation (pp. 1–27). https://edoc.hu-berlin.de/bitstream/handle/18452/19241/Niewoehner2016-Co-laborative-anthropology.pdf?sequence=1&isAllowed=y

Read, M. (2011, April 11). *How to use magnets to mess up your brain*. Gawker. https://www.gawker.com/5791070/how-to-use-magnets-to-mess-up-your-brain

*Regulations for Mingler Scholarship*. (n.d.). https://akademievankunsten.nl/

Roepstorff, A. (2001). Brains in scanners: An Umwelt of cognitive neuroscience. *Semiotica, 2001*(134).

Rogers, H. S., & Halpern, M. K. (2021, forthcoming). Introduction: The past, present, and future of art, science, and technology studies. In H. S. Rogers, M. K. Halpern, D. Hannah, & K. de Ridder-Vignone (Eds.), *Routledge handbook of art, science, and technology studies* (n.p.) Routledge.

Rose, S., & Rose, H. (2016). *Can neuroscience change our minds?*. Wiley-Blackwell.

Salter, C., Burri, R. V., & Dumit, J. (2016). Art, design, performance. In U. Felt, R. Fouché, C. A. Miller, & L. Smith-Doerr (Eds.), *The handbook of science and technology studies* (pp. 139–168). MIT Press.

Slaby, J., & Choudhury, S. (2017). Proposal for a critical neuroscience. In M. Meloni, J. Cromby, D. Fitzgerald, & S. Lloyd (Eds.), *The Palgrave handbook of biology and society* (pp. 341–370). Palgrave Macmillan.

The Editors of Nature. (2021). Collaborations with artists go beyond communicating the science. *Nature, 590*(7847), 528–528. https://doi.org/10.1038/d41586-021-00469-2

Van Lente, H. (2000). Forceful futures: From promise to requirement. In N. Brown, B. Rappert, & A. Webster (Eds.), *Contested futures: A sociology of prospective techno-science* (pp. 43–63). Ashgate.

Wisnioski, M., & Zacharias, K. (2014, May 15). Sandbox infrastructure: Field notes from the arts research boom. *ARPA Journal*. http://www.arpajournal.net/we-are-test-subjects-2/

# The Artificial Womb: Speculative Design Meets the Sociotechnical History of Reproductive Labor

Patricia de Vries

In 2019, researchers from Máxima Medical Center in Eindhoven and scientists at the Eindhoven University of Technology, the Netherlands, received a €2.9 million grant from the EU-program Horizon 2020 to develop an artificial womb. By 2025, they will in all likelihood have developed a prototype. Announcing its development, an image of an artificial womb prototype went viral in more than 3 million online search results. The image of the prototype was taken during the acclaimed Dutch Design Week in 2018, where it was first presented as design-for-debate by Professor Guid Oei of the Máxima Medical Centre. Hendrik-Jan Grievink and Lisa Mandemaker, designers affiliated with the Dutch Amsterdam-based studio Next Nature Network, designed the prototype in collaboration with the Eindhoven team.

Next Nature Network also worked together with the team in Eindhoven to organize a design-fiction exhibition titled *Reprodutopia* (2019). The

P. de Vries (✉)
Lectoraat Art & Public Space (LAPS), Gerrit Rietveld Academie,
Amsterdam, The Netherlands
e-mail: patricia.devries@rietveldacademie.nl

© The Author(s) 2023
K. Bijsterveld, A. Swinnen (eds.), *Interdisciplinarity in the Scholarly Life Cycle*,
https://doi.org/10.1007/978-3-031-11108-2_13

exhibition, curated by Next Nature Network, opened in October 2019 at Droog Gallery in Amsterdam. By way of speculative design objects, *Reprodutopia* imagined future scenarios around the artificial womb that were meant to spark debates about the ways in which a fully functioning artificial womb may alter our attitude toward reproduction, relationships, and love in the twenty-first century.

In this chapter, I focus on the installation of the prototype of the artificial womb at *Reprodutopia.* I take its speculative design as entry point into an exploration of the cultural and medical histories braided around the future scenarios of the artificial womb. I will show that the speculative design of the artificial womb—including the accompanying design objects on display at *Reprodutopia*—echoes several past experiments and imaginations of women's bodies, notably pertaining to the uterus, as sites of ongoing socio-cultural and biological contestation.

Analyzing the prospects of the artificial womb—itself already an interdisciplinary undertaking—in the setting of a futuristic speculative design exhibition, I aim to show that speculative art has greater leeway than academic writing to transgress limits, to max-out ideas, and to use imagination to let us experience and imagine things that do not yet exist or seem to be impossible. With this approach, I endorse trends in scholarship aimed at showing how stories and imaginaries are intrinsic to doing science (e.g., Jasanoff & Kim, 2009; McKittrick, 2021; Felt, 2014). Reproductive technologies have the potential to alter the meaning making by and behavior of their users profoundly, which calls for interdisciplinary involvement in reflecting on such impact through speculative design (Verbeek, 2006).

Speculative design itself, however, also deserves scrutiny, as it draws on, articulates, and materializes, as I will show, *particular* sociotechnical histories. Knowledge of such histories enables us to understand the direction of the speculative design under study. What I aim to make explicit in this chapter is what it takes in terms of interdisciplinary work to show the variety of histories involved. Although I will discuss one project that resulted from the interdisciplinary collaboration between artists and scientists, namely *Reprodutopia*, I need other forms of interdisciplinarity to analyze the messages implied in the speculative design of the exhibition.

# Speculative Art and Emerging Reproductive Technologies

At first glance, design may seem to carry little weight in discussions on the future of reproduction technologies. Yet, as claimed by Marshall McLuhan (2003) in *Understanding Media: The Extensions of Man*, reflections on new technologies require an artistic eye: "The serious artist is the only person able to encounter technology with impunity, just because he [sic] is an expert aware of the changes in sense perception" (p. 31). Such a statement may have an overly romantic ring to it, and I do not agree with McLuhan's idea that *only* artists can understand the days and ages we live in. However, as Oscar Wilde once famously wrote, "Life imitates Art far more than Art imitates Life" (Wilde, 1899, p.17). Artistic imaginaries play a vital role in the development of new technologies, and our ideas about emerging technologies are strongly influenced by the stories we tell.

Reflection on the possible impacts of emerging reproductive technologies and on the future embedding of these technologies in our societies involves a complex task. Scholars in science and technology studies (STS) have pointed out that professionals find it difficult to imagine unknown futures (Felt et al., 2009). Doing so requires that we make tangible and palpable what is usually intangible and impalpable. Yet, art practices can help us to do just that. Artists come equipped with a toolkit that can generate ways of relating to new technologies, showing a wider repertoire of responses—such as affective reasoning and embodied knowledge (Roeser et al., 2018; Cuhls & Daheim, 2017)—than those merely focused on quantifiable impacts, abstract reasoning, and the exchange of rational arguments.

Artworks, in fact, are sites of meaning through which ideas and stories about emerging technologies are organized, shaped, stretched, and circulated. Artistic representations of reproductive technologies—be it in film, literature, or speculative art practices—can show us how people relate to these technologies. Art practices contribute to the materialization of shared norms and values, while the future of reproduction and reproductive technologies has meanwhile occupied the mind of artists for decades—think of Lilith in Octavia Butler's novel *Dawn* (1987), the hatchery of Aldous Huxley's novel *Brave New World* (1932), or the fetus fields of sci-fi movie *The Matrix* (Lana Wachowski and Lilly Wachowski, 1999). Art may thus help us to move beyond specific limits of our views and conceptualizations.

What speculative art and design add to the sciences are tools and resources to reflect on social norms and values, anxieties, expectations, and desires braided around emerging technologies. Anthony Dunne and Fiona Raby (2013), for example, describe their speculative exploration of future scenarios as a way of materializing critical thought to engage people in thinking about possible futures. Seen in this way, speculative design is a form of philosophy of technology that hails potential futures in the present to (re)think the social and cultural implications of emerging technologies. Speculative art practices, then, can be effective in questioning the set of narratives and subject positions triggered by developments in reproduction technologies. In Donna Haraway's wonderful words: "It matters what matters we use to think other matters with; it matters what stories we tell to tell other stories with; it matters what knots knot knots, what thoughts think thoughts, what descriptions describe descriptions, what ties tie ties. It matters what stories make worlds, what worlds make stories" (2016a, p. 4). Speculative art and design can help us think ahead, in an attempt to anticipate unknown futures from different perspectives and positions.

## Welcome to the Future Fertility Clinic

In Next Nature Network's own rendering, as described on its website, *Reprodutopia*'s aim was to explore "the impact of technology on biological reproduction, gender and family" (Next Nature Network, 2019c). The exhibition

> is disguised as a future clinic that presents thought-provoking visions of reproductive technologies by artists and designers … It's time for a much-needed discussion about the way technology radically alters our attitude towards reproduction, gender, relationships and love in the 21st century. If

we are to rewrite the human story, let's make sure it becomes a story that benefits all. (Next Nature Network, 2019c)

The words "future fertility clinic" evoke a sterile, brightly lit space with waiting rooms decorated with minimal design furniture and framed pictures on its walls of happy parents with their offspring. The exhibition space of *Reprodutopia* does much to reinforce this view. Entering the clinic, visitors are greeted by overly friendly employees, dressed in long white coats patched with corporate logos. These fertility officers walk you through the clinic and offer visitors overtly differential, personalized self-help-style advice for designing your prospective reproductive future with your partner(s).

Strolling around the future fertility clinic, it feels as if one has entered a site in between a high-end private medical facility and the Genius Bar at an Apple store. Using humor and exaggeration, *Reprodutopia* offers a variety of speculative design objects on the topic of artificial reproduction. Take, for example, the strap-on uterus that visitors can try on. This design piece resembles a wraparound baby sling (or belt). When wrapped around one's body with the straps tied, it supports the artificial womb from a carer's body. The idea is that in the future, parents and caretakers can freely share the carrying of the portable artificial womb.

To be sure, the prototype exhibited at *Reprodutopia* will not be used to help grow premature babies. The forthcoming Eindhoven prototype will be closer to a so-called biobag container. It will surround an extremely premature baby with fluids and delivers oxygen and nutrients through an artificial placenta that connects to the baby's umbilical cord. This yet-to-be-developed prototype will provide premature babies with an environment that simulates physiological conditions, to mitigate the often-chronic health issues premature babies are likely to suffer from in their lives. Due to a combination of organ immaturity and iatrogenic injuries, extreme prematurity is the leading cause of neonatal mortality and morbidity. In the US alone, over one-third of all infant deaths and one-half of cerebral palsy cases are attributed to prematurity (Partridge et al., 2017). Extending gestation artificially could, so is the expectation, reduce the risk of mortality, disability, and chronic illness associated with extreme premature birth. Scientists working in this field consider birthing a deadly affair as well. As the surrogacy researcher Lewis writes, "hundreds of thousands of humans die because of their pregnancies every year" (2019, p. 1). In the US, about 1000 people die during childbirth and another 65,000 come dangerously close to dying. Unsurprisingly, safer gestation has always been the privilege of the white and wealthy. The medical hope is that developments in the artificialization of reproduction could lower the deadly risks of pregnancy and birthing, as well as prevent miscarriages and maternal mortality.

## Anticipating the Future

Speculative art and design efforts comprise future visions, grounded in a shared present, that enable us to (re)imagine social realities, offering insights into how the world might be, or be made differently, in the future (Mann, 2018). The mandate of speculative design can be to spark debate and, in the case of *Reprodutopia*, to imagine future sociotechnical scenarios to raise questions about the interrelated ethical issues and social consequences as we can conceive them today.

The centerpiece of *Reprodutopia* was Next Nature Network's artificial womb prototype, the one presented during the Dutch Design Week in 2018 by Professor Oei, one of the lead medical scientists of the Máxima Medical Centre in Eindhoven. The prototype, which filled an entire exhibition room to the brim, consisted of a collection of five synthetic air-filled spheres, the size of office ball-chairs, suspended from the ceiling. In the adjoining exhibition room, one entered the reproductive clinic. Upon entering, visitors were immediately accosted by one of the employers of the clinic eager to show them the items for sale.

It seems that the future of medical fertility clinics is fertile ground for merchandise and commerce-driven ritualization, as evidenced by the *Virgin Parent Ring* (Next Nature Network, 2019b) and *Lab Romanticism* (Next Nature Network, 2019a), two design pieces in the exhibition. In theory, artificial reproduction would allow for immaculate conception—virgins could grow a child in the artificial womb, like a present-day Mother Mary and Joseph. This possibility inspired the design of the *Virgin Parent Ring*. The two rings, which have the words "virgin parent" engraved on their outside, will be available for purchase at the future clinic and may be exchanged between two parents. *Lab Romanticism,* in turn, has been designed to provide prospective parents with a selection of mindful romantic rituals to give some color and heart to the "detached" medical procedure of artificial reproduction and the sterility of the future reproduction clinic.

Like the *Virgin Parent Ring,* it "updates" long-standing, semi-religious, ritualistic, and romantic practices. Tasks include lighting candles, raking sand in a mini Zen garden, and exchanging rings, to offer future parents a semblance of romance. Through these two design pieces, *Reprodutopia* appeared to allude to the perpetuation and pervasion of commodification and marketization into the future artificial reproduction clinic, as well as to the persistent conservatism of the romantic imperative and coupledom.

Fertility and reproduction, as we all know, have co-determined the position of women since the fifteenth century. The carrying and birthing of children have been biologically assigned to women. Within capitalist, heteronormative, and patriarchal cultures, childcare and child-rearing have historically been imposed on women as unwaged labor. And the uneven division of labor in gestation, birthing, and parental care continues to be an important factor in the imbalance of the sexes.

Therefore, the development of this potentially disruptive humanoid organ obviously evokes important questions and horizons regarding the interrelations between reproduction, gender, and parenting. What are the affordances of reproduction without pregnancy and birthing? What might change between the sexes through artificial reproduction? What are the possible implications of the separation of sex and reproduction? Who reproduces, with whom (if anyone), and who takes care of the child? What might change in gendered parenting roles? Could this innovation lead to reproductive parity? And who benefits from these futures?

The *Parenting Kit* (2018), another Next Nature Network design, tacitly touches on the possibility of multiple-parent reproduction. Its design resembles a hybrid of an online DNA test-set and Microsoft software package. The description urges visitors to imagine a future in which anyone of us could send off a skin sample to a futuristic lab, and, through a process called "in vitro gametogenesis," have these skin cells transformed into both sperm and eggs. In this way, it would become possible to fertilize ourselves and create a baby on our own. The *Parenting Kit* enables us to have a child without a genetic partner, with a partner of the same sex, or to have a child with a group of individuals who together contribute 100% of the genetic material required for in vitro gametogenesis. The *Kit*, the description text concedes, allows for different forms of parenthood to emerge and be fostered. In the future fertility clinic, parenthood is available in Mono, Duo, and Poly versions. The *Kit* will further enable tinkering with and assembling a curated cocktail of genetic material to gestate offspring in the artificial womb. Here, *Reprodutopia* tacitly touches on the possibility of multiple-parent reproduction, a subject hardly discussed in academic scholarship in relation to reproductive technologies and artificial reproduction.

## Echoes of the Past in Speculative Future Visions

Stories we *do* tell about the future of reproduction—whether imagined in the form of design-fiction, narrativized in science-fiction, or represented in films—are situated and embedded. They have histories, be it cultural histories, histories of science, technology and medicine, or histories of ideas, to mention just a few options. It is not at all self-evident, however, which of these histories are relevant for a critical analysis of what speculative design performs. Speculative design may be a form of philosophy of technology, while showing on which histories it draws will depend on the analyst's literacy in historical accounts and the analyst's conceptual ability to identify specific similarities with or particular absences in past practices. Below, I will present the similarities and absences *I* noticed.

The first thing that struck me when observing the prototype exhibited at *Reprodutopia* is that the bunch of air-filled, flesh-colored balls look eerily like loitering testicles. It is of importance—and tragicomic—that the prototype has more likeness to dangling testicles than to the shape of, say, a womb. The "absence" of the *form* of the womb in this setting testifies of a long history of power relations at stake in artificial womb research and design. The lead medical scientists of the incubator-cum-artificial-womb project at Eindhoven are an all-male cast. Mentioning their sex is noteworthy here. The development of an artificial womb consists of situated and embodied acts. "Situated knowledges," a term coined by Haraway in 1988, describes the interrelated and inseparable planes of ontology, epistemology, politics, and ethics. All knowledges, Haraway argued, are situated knowledges. Science is a doing, and this doing is done by bodies. Bodies are marked, and their marking is always determined by their role in "scientific and technological, late-industrial, militarized, racist, and male-dominant societies" (Haraway, 1988, p. 581). By implication, then, the development of the artificial womb, as embodied and situated acts of knowledges, is inevitably overburdened by existing power relations.

The second thing I noticed is that the artificial womb in Eindhoven represents steps in the increasing artificialization of biological reproduction. Its nomen, "artificial womb," has breathed new life into the age-old vision of the growth of a human embryo outside the body—ectogenesis, from the Greek "ecto," outer, and "genesis," birth. The hanging spheres evoke images of the alchemical homunculus—the mother of all innovations: artificially created life. And not without reason. The aspiration to design an artificial womb, or a womb replica if you like, is an age-old dream in the history of medical science. Its history can be traced back to early automata when medieval alchemists huddled around glass containers trying to conjure up miniature men in tiny bottles. Some medieval alchemists liked to believe that a homunculus could have superior powers, if not become a morally and spiritually better version of the human. Speculative future visions of reproductive technology hark back to these earlier fantasy visions of the homunculus.

There are, of course, different interpretations of the story of the homunculus and how it relates to the development of the artificial womb. One may argue that the homunculus actually represented the erasure of women from sexual reproduction by technology: men creating beings without female involvement. The same can be said to apply to the artificial womb. One may argue that the prospect of artificial reproduction will liberate women from the health risks involved in carrying and birthing babies, re-assign agency to women, free them from the constraints of their bodies and so-called biological clocks, and fundamentally change the labor relations and gender divisions in society. Others may argue that although the artificial womb is likely to open possibilities for non-normative forms of parenting, while also mitigating health risks associated with prematurity, gestation, and birthing, large-scale adoption of such womb might as well exacerbate socio-economic inequalities if access to artificial reproduction is not equally distributed. In this respect, Shulamith Firestone, a leading feminist thinker and major proponent of artificial womb technology, wrote in *The Dialectic of Sex* (1970) that "[i]n the hands of the present society there is no doubt that the machine could be used—is being used—to intensify the apparatus of repression and to increase established powers" (p. 193). Firestone actually feared that artificial reproduction could be used to repress women when conditions of gender inequality would remain unchanged in society.

The yet-to-be-developed artificial womb is to function as a new type of "incubator," another similarity I identified, meant to intervene in the high number of premature baby deaths every year. Jeffrey P. Baker's *The Machine in the Nursery* (1996) offers a case study of the development of the incubator from its origins in the Paris maternity hospital between the years 1880 and 1922. In the early days of the incubator, in late 1870s and early 1880s in France, its design was purported to be analogous to the womb: a closed system containing warm fluids and impenetrable to light. In an attempt to lower maternal and infant mortality and premature infants, the French pediatricians Stéphane Tarnier modeled the first infant-warming machine, the *couveuse*, after the chicken incubator. The machine bolstered a nationwide campaign against infant mortality. The device was meant to help mothers and nurses, not to replace them, according to Baker. Moved to the US, the incubator underwent a radical transformation. Baker describes how American pediatricians collaborated with various third parties, such as professional inventors and entrepreneurial physician-inventors (Baker, 1996, p. 67). The American incubator resulting from these collaborations was labeled an "artificial womb" and was meant to replace mothers and nurses (p. 70).

Claire Horn explains in *Psyche* that attempts to create an artificial womb were accompanied by the shared concern of male obstetricians, doctors, and scientists "that mothers themselves, with their [assumed] unsanitary practices, irresponsible behavior and anxious fussing, might pose a danger to their infants—a danger that could be curbed by placing the uterus-incubator firmly in the doctors' hands" (2020). The current developments in the prototyping of an artificial womb by a team of researchers in Eindhoven can be understood in terms of this medical history of the incubator. Although the artificial womb is developed to lower the risks of extreme prematurity—the leading cause of neonatal mortality and morbidity—the incubator was developed in part because American doctors and scientists thought that the number of premature infants and the level of neonatal mortality could be reduced by separating women from their babies.

Future visions of the artificial womb thus mirror broader social values, systems, and histories. Take, for example, the seemingly convenient and benign portable womb-sling exhibited at *Reprodutopia*, allowing expectant parents to share the weight of carrying the fetus. This sling needs to be situated in the broader history of the uterus as a site of capitalist labor exploitation. The uterus underlies a major historical shift in Western societies. In her seminal book, *Caliban and the Witch: Women, The Body and Primitive Accumulation* (2004), Silvia Federici exhaustively documents how control over the uterus was critical to the foundation of capitalism. Key drivers of the development of capitalism in Europe were colonization, the Atlantic slave trade, the expropriation of the European peasantry from its lands, *and* the repressive control of women's bodies, including unwaged and reproductive work. Federici documents how the primary accumulation of capital implied the development of a new sexual division of labor subjugating women's labor and women's reproductive function to the production of the workforce. She presents the rise of capitalist labor in Europe as a development that fundamentally undermined the position of women in society. According to Federici, creating surplus value in capitalist economies became possible only because of the forced labor of enslaved workers, and the unpaid housework and reproductive labor of women confined to the domestic sphere and excluded from waged work. Forced and unpaid labor, including unpaid reproductive labor, created and sustained the conditions for the production of value, she argues. Such labor fueled the construction of a new patriarchal order, based on women's subordination to men. Finally, and crucially, it also resulted in the mechanization of women's bodies as machines for the production of new workers, which included the criminalization of abortion, and the decriminalization of rape of proletarian women. Federici describes the various ways in which a concerted effort was made on the part of the church and the state to undermine class solidarity and divide the emerging working class along gender lines. Women, Federici argues, were to produce labor-power for the farms and workshops and cannon fodder for the imperial wars.

## Conclusion

To this day, for too many women, fertility and motherhood mean risking their bodies, their careers, and sometimes even their lives, while being constantly scrutinized, judged. The cultural theorist Valeria Graziano makes a poignant argument, on Facebook in 2018, about motherhood that is worth quoting at length:

> For too many women, motherhood is not a choice as they are pressured into it as the only social role available to them. For too many women, motherhood is a choice, yes, but of giving up on other practices, studying, working, creating, participating in politics or in the life of their communities, simply because the joys of motherhood are all they are supposed to aspire to while they toil away in the solitary drudgery of domestic labour. For too many women, motherhood can only take the form of sacrificial love, as they exhaust themselves juggling the demands of making a living, of complying with bureaucracy, of confronting the devastating paucity of care provisions. For too many women, because of the demands of making a living, of complying with bureaucracy, and the devastating paucity of care provisions, motherhood is not an option at all.

My critical analysis of *Reprodutopia* shows that its view of motherhood is not speculative enough. It basically involves, as revealed by my interdisciplinary identification of historical similarities and contemporary absences, the projection of the present onto the future, which tethers the future of motherhood to ongoing gendered parenting roles in a neoliberal market as well as a patriarchal culture. If we continue to cling to gendered social formations and neoliberal economic structures, however, this will frustrate the development of alternative futures and different knowledge hierarchies.

In principle, speculative design can foster a critical re-orientation that does neither entirely abandon the historical formations and economic structures of motherhood and gendered parenting roles, nor give in to those structures and formations as a necessary limit on what comes next (Mann, 2018). Going forward, we need more *and* more radically speculative design, as well as more artistic and science-fiction imaginaries of the socio-political and economic futures of reproductive technologies, including their possible implications. We need to feed the public imagination with possible other maternal futures, in an attempt to find "still possible, recuperating pasts, presents and futures" (Haraway, 2016b).

# References

Baker, J. (1996). *The machine in the nursery*. Johns Hopkins University Press.

Cuhls, K., & Daheim, C. (2017). Introduction to the special issue on "Experiencing Futures". *Futures, 100*(86), 92–93.

Dunne, A., & Raby, F. (2013). *Speculative everything: Design, fiction, and social dreaming*. MIT Press.

Federici, S. (2004). *Caliban and the witch: Women, the body and primitive accumulation*. Autonomedia.

Felt, U. (2014). Sociotechnical imaginaries of "the internet," digital health information and the making of citizen patients. In S. Hilgartner, C. Miller, & R. Hagendijk (Eds.), *Science and democracy. Making knowledge and making power in the biosciences and beyond* (pp. 176–197). Routledge.

Felt, U., Fochler, M., Müller, A., & Strassnig, M. (2009). Unruly ethics: On the difficulties of a bottom-up approach to ethics in the field of genomics. *Public Understanding of Science, 18*(3), 354–371.

Firestone, S. (1970). *The dialectic of sex: The case for feminist revolution*. Bantam Book.

Haraway, D. (1988). Situated knowledges: The science question in feminism and the privilege of partial perspective. *Feminist Studies, 14*(3), 575–599.

Haraway, D. J. (2016a). *Staying with the trouble: Making kin in the Chthulucene.* Duke University Press.

Haraway, D. (2016b). Tentacular thinking: Anthropocene, Capitalocene, Chthulucene. *E-flux, 75.* https://www.e-flux.com/journal/75/67125/tentacular-thinking-anthropocene-capitalocene-chthulucene/

Horn, C. (2020). The history of the incubator makes a sideshow of mothering. *Psyche.* https://psyche.co/ideas/the-history-of-the-incubator-makes-a-sideshow-of-mothering

Jasanoff, S., & Kim, S. H. (2009). Containing the atom: Sociotechnical imaginaries and nuclear power in the United States and South Korea. *Minerva, 47*(2). https://doi.org/10.1007/s11024-009-9124-4

Lewis, S. (2019). Full surrogacy now. *E-flux, 99.* http://worker01.e-flux.com/pdf/article_261641.pdf

Mann, J. L. (2018). Pessimistic futurism: Survival and reproduction in Octavia Butler's Dawn. *Feminist Theory, 19*(1), 61–76. https://doi.org/10.1177/1464700117742874

McKittrick, K. (2021). *Dear science and other stories.* Duke University Press.

McLuhan, M. (2003). *Understanding media: The extensions of man,* edited by W. Terrence Gordon. Gingko Press.

Next Nature Network. (2018). *Parenting Kit* [design object]. Droog Gallery, Amsterdam, The Netherlands.

Next Nature Network. (2019a). *Lab Romanticism* [design object]. Droog Gallery, Amsterdam, The Netherlands.

Next Nature Network. (2019b). *Virgin Parent Ring* [design object]. Droog Gallery, Amsterdam, The Netherlands.

Next Nature Network. (2019c). *Reprodutopia.* https://nextnature.net/projects/reprodutopia

Partridge, E., Davey, M., Hornick, M., McGovern, P., Mejaddam, A., Vrecenac, J., Mesas-Burgos, C., Olive, A., Caskey, R., Weiland, T., Han, J., Schupper, A., Connelly, J., Dysart, K., Rychik, J., Hendrick, H., Peranteau, W., & Flake, A. (2017). An extra-uterine system to physiologically support the extreme premature lamb. *Nature Communications, 8,* 15112. https://doi.org/10.1038/ncomms15112

*Reprodutopia* [Exhibition]. (2019). Droog Gallery, Amsterdam, The Netherlands. https://nextnature.net/projects/reprodutopia

Roeser, S., Alfano, V., & Nevejan, C. (2018). The role of art in emotional-moral reflection on risky and controversial technologies: The case of BNCI. *Ethical Theory and Moral Practice, 21,* 275–289. https://doi.org/10.1007/s10677-018-9878-6

Verbeek, P. (2006). Materializing morality: Design ethics and technological mediation. *Science, Technology, & Human Values, 31*(3), 361–380. http://www.jstor.org/stable/29733944

Wilde, O. (1905 [1899]). The decay of lying. In *Intentions*. Brentano. http://virgil.org/dswo/courses/novel/wilde-lying.pdf

# Doing Collaborative Research on Symphonic Orchestra Audiences: Interventionist Ethnography of Music Practices

Peter Peters, Ties van de Werff, Imogen Eve, and Jos Roeden

P. Peters (✉)
Department of Philosophy, Faculty of Arts and Social Sciences,
Maastricht University, Maastricht, The Netherlands
e-mail: p.peters@maastrichtuniversity.nl

T. van de Werff
Lectoraat Autonomie en Openbaarheid in de Kunsten, Zuyd University of
Applied Sciences, Maastricht, The Netherlands
e-mail: ties.vandewerff@zuyd.nl

I. Eve
Master Student Leiden University, Leiden, The Netherlands

J. Roeden
Department of Programming and Planning, philharmonie zuidnederland,
Maastricht, The Netherlands
e-mail: jos.roeden@philharmoniezuidnederland.nl

© The Author(s) 2023                                                                      **253**
K. Bijsterveld, A. Swinnen (eds.), *Interdisciplinarity in the Scholarly Life Cycle*,
https://doi.org/10.1007/978-3-031-11108-2_14

# Introduction

In the social sciences and humanities, collaboration across disciplines, including the arts, increasingly features as an extension of the repertoire of conventional research methods. As a programmatic ideal, it is thought to address challenges that higher education institutes and universities face in circulating and valorizing the knowledge they produce. As Georgina Born and Andrew Barry (2014) have argued, the current prominence of collaboration across disciplinary boundaries is linked to changing relations between science, technology, and society, an increasing need for the accountability and reflexivity of research agendas, and the claim that innovation in knowledge societies depends on interdisciplinary collaboration (2014, p. 1). A similar move toward forms of interdisciplinary and transdisciplinary collaboration can be seen in the arts. Today, art worlds are often heterogeneous and include a broad range of actors and audiences. In contrast to traditional artistic production in an art academic and primarily crafts-based environment, artworks are now often created in academic, social, and economic settings that are institutionally diverse.

In our chapter, we will focus on collaborative research carried out by the Maastricht Centre for the Innovation of Classical Music (MCICM).[1] This inter- and transdisciplinary collaboration between an orchestra, a higher arts education institute, and a university situated in the South of the Netherlands started from sharing a problem: how can symphonic orchestras shape new futures through innovating their practices? Each of the partners has a stake of its own in addressing this problem. Whereas the orchestra hopes to attract new audiences and strengthen its public presence, the conservatory aims to update its curricula and the academic researchers are interested in orchestral music as a major practice of cultural transmission. Reflecting on our work in the MCICM in recent years, we are interested in how the initial idea of setting up the orchestra as a laboratory for practice-based and artistic research on new concert formats and audience participation developed into an everyday reality of

---

[1] The partners in the MCICM are Maastricht University (UM), the South Netherlands Philharmonic (philharmonie zuidnederland), and Zuyd University for Applied Sciences (Zuyd), which houses the Conservatorium Maastricht. The MCICM is co-funded by the three partners and by the Province of Limburg, the Netherlands.

collaborative learning. How did this collaboration play out in practice? What was successful and why? And what, perhaps, proved to be less effective?

To answer these questions, we discuss the NWO-SIA funded Artful Participation project (2017–2021) as an example of interventionist ethnographic research on symphonic music audiences (see Artful Participation, 2021, December 1). The design of this project, carried out by the partners of the MCICM, reflects the aim of the 2016 Smart Culture call for proposals that "in the area of arts and culture, fundamental and practice-oriented research can enhance each other" (Call, 2016, p. 2). This call echoed the claim that collaborative research can lead to innovation, in this case of the "ecosystem of the creative sector" (p. 2). Following this strategy, the Artful Participation project sought to combine strategic research into reasons for the declining interest in symphonic music with embedded research aimed at innovating this practice in artistically relevant ways. The collaborative research took place in a series of specifically designed experiments with audience participation in symphonic events. Our reflection on these experiments resulted in a learning model that aspires to help symphonic orchestras to innovate their practices, in particular when it comes to audience participation.

Elaborating on an experiment called The People's Salon, we will show how the practical work to make the experiments happen can be traced through the many conversations that shaped the collaborative process. To understand why orchestras focus on audiences when innovating their practices, we first provide an overview of recent developments in the symphonic sector. Next, we present several basic ideas behind the research design of the Artful Participation project and The People's Salon experiment. Reflecting on vignettes from our fieldwork through the lens of Richard Sennett's work (2012) on the rituals, pleasures, and politics of cooperation, we draw conclusions about the role of conversations in collaborative research. Following Sennett, we use the terms 'collaboration' and 'cooperation' interchangeably, as synonyms, even though we realize that they can have different meanings and connotations in various contexts.

# Innovating Symphonic Music Practice

Symphonic orchestras in the Western world are faced with challenges that affect their status as cultural institutions embodying a living classical music tradition. Even when orchestras perform contemporary music, many of them seem to function more like museums. In a heterogeneous musical landscape, most of them have been focusing on a canon of symphonic works from the late eighteenth to the early twentieth century. This development coincided with the emergence of debates on the cultural and social relevance of symphonic orchestras, as reflected in the arguments for their funding. Starting in the 1980s, neoliberal cultural policies increasingly questioned the role of the government as the main funder of cultural institutions. In recent decades in the Netherlands, this gave rise to a long series of budget cuts, forcing several symphonic orchestras to merge, while others in fact ceased to exist. Today's market imperative introduces a paradox of legitimation: symphonic orchestras need to be funded because they are important, but if they are so important, why are there not enough people prepared to pay the full price of their tickets? Key criteria for funding continue to be linked to the need to attract new audiences and to create connections in a rapidly changing world (Ministerie van OCW, 2011, p. 37; Ministerie van OCW, 2013, p. 1; Raad voor Cultuur, 2014, pp. 41–43).

In response to these various challenges, symphony orchestras have tried to critically reevaluate and innovate their practices (Idema, 2012). Today, many orchestras engage with local communities or play music in classrooms, thus finding other sites to perform beyond the concert hall (e.g., the Scottish Nevis Ensemble, www.nevisensemble.org). Concertgoers are encouraged to read about the music they hear in real time on their smartphones with apps such as Wolfgang (www.wolfgangapp.nl/). More and more concerts can be attended through livestreaming, as if performed in a digital concert hall, where, as it is put on the website of the Berlin Philharmonic, "we play just for you" (www.digitalconcerthall.com/en/home). Other orchestras, such as the Dutch Pynarello, have tried to break with concert conventions by performing without scores (www.pynarello.com), while Ensemble Modern gave the audience a role as artist in the concert process (www.ensemble-modern.com/de/projekte/aktuell/connect-2016), aiming to re-explore the relationship between composer, musician, and audience and to enable the audience to participate in concerts more actively (Toelle & Sloboda, 2019). All of these innovations have contributed to changing the ways in which audiences participate in symphonic concerts (Peters, 2019).

## Participating in What?

In the current symphonic practice, audiences are mostly conceived as listener, consumer, or amateur. In the Artful Participation project, we experimented with ways to change these roles into maker, citizen, or expert, thus actively involving audiences in programming, co-organizing, and assessing symphonic music concerts. Our research design elaborated on recent work on musicology and music sociology that aims to close the gap between page and stage (Cook, 2014), between the musical work and the practical work that needs to be done to make music happen. In line with this goal, we understand music in the making as a social, material, and situated practice (Small, 1998; Born, 2010; Hennion, 2015). Drawing on this practice approach, we studied empirically how audiences participate in music performances, using insights from fields such as audience research in the performing arts (Burland & Pitts, 2016), but also from science and technology studies (STS). A central insight from STS research on music and its instruments is that engaging users in the development of an innovation is key to its successful adoption (e.g., Pinch & Bijsterveld, 2003). STS researchers have also argued that every innovation involves prescriptive choices, often implicit. This is certainly the case in the normatively charged practice of symphonic music, where aesthetic norms are constitutive of the way concerts have been organized since the early nineteenth century (Bonds, 2014).

Symphonic practice revolves around the performance of musical works. In what she calls the "Beethoven paradigm," Lydia Goehr (1992) argues that the work concept regulated how composers notated their music, how performers were expected to be true to the score to give authentic performances, and how audiences listened in silence to hear the beauty of the work itself (see also Smithuijsen, 2001). In the Beethoven paradigm, music and aesthetic experience in general are abstract because they derive from a realm of the beautiful that is timeless. The aesthetic experience must focus on the artwork as such, while refraining from non-aesthetical aspects such as goal, function, and situation. By disregarding the original context of life, the music becomes visible as a pure work of art. Historically, this process of abstraction has also created places solely dedicated to art, such as the museum, the theater, and the concert hall. Today's classical music practice reflects many of the aesthetic assumptions of the Beethoven paradigm and its work-centeredness.

In our approach of the symphonic practice, the musical work cannot be isolated from the conditions under which it is presented. In fact, it can only exist in its relations to the lifeworld. Drawing on the ideas of philosophers such as Hans-Georg Gadamer (1960/1989) and John Dewey (1934/2005), we are interested in how presenting musical works under different conditions leads to productive variations. Performing symphonic music, then, is a matter not of replicating earlier acts of presenting it, as argued by Gadamer, but of creating new renderings that keep the future identity and continuity of the musical work open. Dewey follows a similar line of reasoning against what he calls the "museum conception of art" (1934/2005, p. 4). Instead of understanding works of art in their external and physical existence, detached from the actual life-experience from which they emerge and in which they have consequences, we should show how the aesthetic experience is rooted in everyday experiences. We constitute the work of art through our interactions with it, using past experiences to provide it with new meanings. In our project, we radically conceptualized a musical work as an entity that has to be continually performed and worked upon to exist at all (Peters, 2019).

Following this reasoning, we understand musical performance as contributing to the life of a composition by extending its tradition in new ways. This means, first, that the performance of a musical work can be seen as extending its trajectory through actualizing it, or, in other words, giving it meaning in contemporary social, cultural, and technological contexts. Why this performance, why here, why now? Second, a musical performance has to be organized as a material and situated event that involves the work of many. In our project, we started from the assumption that symphonic music audiences can actively contribute to the musical performance that brings music into existence. And finally, a musical work only exists as it is given value in the here and now. Conventional classical music practices seek to render excellent performances of musical works, the aesthetic value of which is considered as a given. Understanding this value as an audience member requires the cultural capital that comes from *Bildung* in the arts and a musical education. In our project, we were interested in ways to experience not-given, emerging values of symphonic music.

## Creating The People's Salon

The experiments in the Artful Participation project were set up as interventionist ethnographies. These were aimed at creating events that would generate 'living' artistic experiences (Marres, 2012) as well as knowledge through collaborative making and reflecting. Actually, designing and performing concerts together with philharmonie zuidnederland was a way to learn about audience participation through observation and intervention. The experimental concerts were designed by Imogen Eve, the musician-researcher in the project, and co-organized by Ties van de Werff, responsible for the learning model that is one of the outcomes of the project. Peter Peters coordinated the project on the side of the researchers as the project's principal investigator, together with Jos Roeden on the side of the orchestra being responsible for the orchestra's programming (Artful Participation, 2021, December 1).

The People's Salon started with a 'mood board' created by Imogen Eve:

> This evening demonstrates how a community can take a shared responsibility and ownership of classical music through programming and hosting their favourite repertoire. The salon further reflects on the long history of the music salon itself, which has been a medium for classical music and the "meeting of minds." Possibly we can invite also new audience groups to partake in these salons (i.e., young people, non-regularly concert goers, etc.). (Eve, 2019, p. 1)

The project design focused on collecting stories and memories that the Friends of the Philharmonic, an association of audience members sponsoring the orchestra, shared through interviews and focus groups, where each individual was supposed to describe how a particular piece of music is valuable for them. The repertoire should consist of pieces that triggered memories and evoked shared values contained in these stories. The performance was to be held on one night as a promenade salon: the audience and musicians would gather in a room for the first piece and move on to a different room, a different piece, a different memory, and a new 'value/issue' to share. The performance was designed as an immersive experience for the audience. The repertoire had to be varied in terms of ensemble sizes, while the venue would also have to be big enough to contain a small-scale orchestra and yet intimate enough to have conjoining rooms to small spaces: "The aesthetic of this production is incredibly important and therefore we will need to ensure the space and the time for designing the space" (Eve, 2019, p. 2).

In the Fall of 2019, Van de Werff and Veerle Spronck, the PhD candidate in the project responsible for the academic research on audience participation in symphonic practices, conducted individual interviews with Friends of the Philharmonic and Van de Werff organized two focus groups. A group of twelve Friends participated in the focus groups, as well as members from the project team and Jos Roeden as the orchestra's programmer. The first meeting of the focus group gave the Friends the opportunity to share their stories and relate them to specific compositions. A longlist of works that might be performed was drafted by the research team. During a second focus group meeting, a smaller group of Friends finalized the repertoire for the evening, with the research team and Jos Roeden in an advisory role. It turned out that many of the compositions mentioned were either instrumental solo pieces or chamber music. Orchestral works selected by the Friends required a big orchestra which was not foreseen in the orchestra's schedule. The solution found was to choose smaller works and, in the case of one piece for a large orchestra, *Sheherazade* by Rimsky-Korsakov, to make a new adaptation (by orchestra musician Roger Niese) for small orchestra. Whereas the program for the concert was developed in close collaboration with the Friends, the choice of the venue was made by the research team. Potential venues in the inner

Fig. 1    The foyer of the AINSI art space, photograph by Peter Peters © Peters

city of Maastricht, the Netherlands, were inspected, some of them with a nineteenth-century atmosphere. In the end, the research team chose a refurbished cement factory on the outskirts of Maastricht called AINSI. The building houses studios for artists and creative entrepreneurs, and it has a large foyer as well as a mid-sized black box theater space. The fauteuils in the foyer added to the salon-like atmosphere (Fig. 1).

The People's Salon concert on January 25, 2020, was attended by 150 Friends, the maximum number allowed in the space. When entering the building, the audience members received a program leaflet that offered a list of the works performed, mainly focusing on the selected stories from the Friends and the ideas behind the experiment. Before the break, piano solo pieces and chamber music were performed, followed after the break by three ensemble and orchestral works. Prior to the performance of the pieces, Friends were interviewed by Han Vogel, timpanist in the orchestra, in a setting with two fauteuils and a bouquet of flowers (Fig. 2). During the

**Fig. 2** Friend of philharmonie zuidnederland being interviewed during The People's Salon, photograph by Jean-Pierre Geusens © Focuss22

break, the Friends were invited to have conservations about the meaning of classical music in their lives, helped by cards with suggestions for talking points. MCICM team members mingled with the audience to make short interviews and to take fieldnotes of their observations. These included the performances and the interactions between the musicians and the Friends.

Adapting the original ideas of The People's Salon to the actual concert situation on the night of performance required a long series of negotiations. Collaboration took the form of months of discussions, scheduling meetings, and making intellectual, artistic, and practical decisions. All this work was documented in plans and working papers, in our fieldnotes, in recordings of the two focus group meetings and the concert evening, in pictures, and video fragments. The heterogeneity of this material—mixing practical communication and reflective observations—offers an insight into what it means to collaborate. Looking back on the event, we realize that what remains is not only this documentation but also the memories of, often undocumented, conversations between everyone involved at various moments and places. Understanding how we collaborated starts with analyzing these conversations.

## Dialectic and Dialogic Conversations

To interpret the conversations recorded in our fieldwork, we draw on Sennett's ideas on the nature of cooperation. In his book *Together* (2012), Sennett defines the concept as "an exchange in which the participants benefit from the encounter" (p. 5) and argues that cooperation is a craft that requires skills. Examples of these skills are "listening well, behaving tactfully, finding points of agreement and managing disagreement, or avoiding frustration in a difficult discussion" (Sennett, 2012, p. 6). A typical situation of cooperation, Sennett argues, is the musical rehearsal. During a rehearsal, musicians do not primarily exchange their individual views on a composition. In fact, through listening well, they become more cooperative creatures (p. 14). Together, they "forensically investigate concrete problems" and work toward a particular moment of collective sound (p. 16). Rehearsing requires rituals and habits as well as the ability to improvise to solve unexpected problems (p. 17).

In his comparison of musical rehearsal to verbal conservation as forms of cooperation, Sennett distinguishes between dialectic and dialogic conversations. The first type of conversation gradually builds up to a synthesis. The goal of the conversation is to find common ground, to come to an agreement, and the cooperative skill involved is to listen to what a person assumes rather than says as a means to detect common ground (Sennett, 2012, p. 19). The second type refers to conversations "that do not resolve [themselves] by finding common ground" (p. 19). Here the goal is mutual understanding while reflecting on the differences between one's positions: "through the process of exchange people may become more aware of their own views and expand their understanding of on another" (p. 19). Sennett compares this type of conversation to a chamber music performance whereby the players do not seem to be on the same page but engage in a sounding dialogue experienced by the audience as more complex and interesting than a polished version of the piece based on agreement (p. 20).

We consider the two forms of conversation that Sennett introduces as ideal types. Dialectical conversation starts from two distinctive positions and leads to agreement in synthesis. This is the type of cooperation of which the value lies in the result. Differences are bridged through a shared commitment to a common goal, which, according to Sennett, assumes sympathy: the willingness to identify with others. In dialogical conversation, the differences are not bridged but taken as a precondition for learning; in making differences explicit and reflecting on them, the conversation itself becomes the goal. This type of cooperation builds on the ability to empathize with others, in other words to try to understand the other's position without giving up one's own: "Curiosity figures more prominently in empathy than in sympathy" (Sennett, 2012, p. 21). Sennett's ideal types help us to analyze four vignettes from our fieldwork, four conversations that made things happen.

The first vignette describes the moment when the technical crew of the orchestra arrived at AINSI the day before the concert to unload the equipment and set the stage. From the research team, Eve and Peters were present:

> I arrive around 10:15 am and take pictures of the orchestra's truck parked in front of the building, feeling excited about us doing this. I look around in the foyer and take some pictures. I meet Imogen—"Hi Peter"—who is there already. The crew of the orchestra is busy lugging things and boxes inside. Werend [the orchestral inspector] has not yet arrived, as I deduce from the comments of the men. In a small room backstage, Imogen shows her sketch of the situation to the men. Immediately, I feel some impatience and irritation on their side. They were not informed of the changes we decided on yesterday, which are a deviation from the project plan they received. It's a slightly tense situation, but Imogen takes charge and tells the men what she envisions: a situation where some of the Friends sit on musicians' chairs, as if joining them. There is a discussion whether this is possible, but eventually the men get to work building the stage. (…)
>
> Werend arrives. Imogen and I explain the situation to him. He reassuringly says all is going to be fine. Werend looks at the set-up for the *Brandenburg concert* and says that the musicians will probably want to stand in a semicircle because they have to be able to see each other. Imogen is not happy. Werend calls the concertmaster, who says—as expected—that

it must be a semicircle. This changes the whole setting into a more traditional orchestral situation, and there is no longer room for the chairs where, as Imogen hoped, the Friends could sit. (Fieldnotes Peters, January 2020 [Original in Dutch, translation PP])

The conversations in this situation were dialectical. In the end, the different positions were bridged in the shared task of setting up the stage in the time that was available. Although the orchestra crew and the orchestral inspector showed sympathy for Eve's set design, in the end, orchestral routines prevailed. Synthesis took here the shape of pragmatic solutions that allowed everyone to reach the goal of being ready for the performance, and the conversations were instrumental to that goal.

On the next day, during the dress rehearsal with the orchestra, Eve shared her design ideas with the musicians, including the colored lighting scheme that changes for each piece. This approach required flexibility and understanding on the part of the musicians, one of them being the pianist who would perform a solo piece and a duo with a violinist. During the solo piece by Mozart, Eve wanted him to be in a yellow spotlight and all the rest of the stage in shadows while he played:

And like a moment pulled from time, a golden lens, a structured spherical 1.5 × 1.5 vignette, he is cut out from another age. A memory. The story that we are forming from remembrance. Pre-War. Post-War. Mid-War. *The nuns at the nursery school had a music box, melodies from Mozart, it was beautiful* [from a story by a Friend] and fusing with yellowed keys, this living music box turns phrases, unlocking synaptic movements, tracing tarnished mechanical cogs like—wrong in the left hand.

'I'm sorry, I'm not sure I can do this.'

The yellow light fizzes as the pianist swivels around, blinking apologetically yellow at me.

'This light is so weird, I mean,' and he laughs, 'I'm looking at my hands but everything is just blurring together.'

I move over towards him and look at the keyboard. Golden brown hands on golden white teeth. My eyes hurt.

'You're right. Like trying to read in the dark.'

'Or underwater,' he laughs.

I sigh.

He swivels around and shrugs, smiles, smiles sideways at me. 'But does it look how you want it?'

I nod, rubbing the back of my neck. 'Yeah. Really beautiful. Really.'

He breathes out and stretches. I can hear the bones in his fingers crack. Then he places his hands on the piano again.

'Well then,' he says, 'Let's give it another go.' (cited from Eve, 2020, pp. 95–96)

We would interpret this conversation as dialectic in that it resulted in a solution to the practical problem: Eve saw the lighting as a way to suggest a different era, the pianist could not see his hands. The fact that the pianist was prepared to give it a try regardless of the difficulties he faced highlights how he sympathized with Eve's ideas and identified with her point of view: does it look how you want it? Although there is mutual understanding and reflection on each other's position, in the end, this conversation lacks the open-endedness of a dialogue. As in the previous vignette, the goal of the conversation was to solve a problem rather than a continued attempt at mutual understanding.

The aim of The People's Salon was to design a situation in which a concert audience, in this case the Friends of philharmonie zuidnederland, could participate by taking responsibility for the program and make a contribution to the actual concert by sharing their stories. Setting up this experimental situation revealed a fine balance between predetermining a certain course of events and leaving room for the unexpected, as became clear during the break when the Friends were invited to talk about the meaning of classical music in their lives:

How different is the kind of participation I now witness, during the break of the concert! We had hoped that the concert program—which included small ensembles and short interviews with Friends about their personal stories and memories—would trigger conversation among the audiences present, about classical music. To encourage audiences to talk about classical music, I had put little cards on the table, with some questions that could start a conversation. But now, when strolling around the foyer during the interval, I hear that people are talking about a lot of things but not about classical music. (Fieldnotes Van de Werff, January 2020)

Understanding the pitfalls of doing participatory experiments occurred when Van de Werff realized that planning a discussion through talking points goes against the idea that good conversations follow their own, improvised course. In Sennett's terms, his approach was dialectical in that its linear structure—with a clear idea of what the cooperation between researcher and audience entailed, and how its outcome of the situation should be—did not account for the complexities which might develop in this cooperation (Sennett, 2012, p. 26–27).

In the case of The People's Salon, cooperation between researchers and orchestra meant that traditional roles and responsibilities were exchanged to a certain degree. As researchers we took charge of the organization of a concert evening, taking over artistic and organizational tasks from the orchestra. For orchestra musicians and staff, the project meant that they were invited to observe and evaluate the concert as an experiment aimed at learning about their interaction with the audience. More than a year after the concert, Roeden and Peters looked back on how they remembered the evening in a long conversation, from which the following vignette is a sample:

Roeden: I deliberately did not sit in between the audience. I stood at the side. I tried to focus on the interaction between the orchestra playing and the audience in order to be able to look the audience in the eye and see what originated there, what happened there.

Peters: And what did you see?

Roeden: The interaction, the disappearance of that anonymity that normally characterizes the division of labor between orchestra and audience. Producing something as a collective, performing something for each other. And enjoying it on both sides.

Peters: I think so too. It was very beautiful. You also could see that making music in a broader sense—I am not talking about producing the sound, but music as an experience—actually became a shared responsibility of the audience and musicians.

Roeden: Yes, yes. (Conversation between Roeden and Peters, May 2021)

This conversation more than a year after the concert is dialogic. Throughout the entire project, one of the main challenges was to find out how to work together from very different starting positions—as researchers and as orchestra staff and musicians. Often, these positions were taken for granted. Sometimes, we managed to truly empathize with each other, which is, as Sennett claims, a more demanding exercise, "at least in listening: the listener has to get outside him- or herself" (2012, p. 21). Roeden empathized with the position of the researcher by taking an observer's point of view. He was curious not only to see how the audience would react but also to find out what would happen at a concert in which he shared the artistic responsibility with the researchers and the Friends. In their conversation, Roeden and Peters reflected on the differences in their perspectives and how they learned from these differences without transcending them in a synthesis or common ground. They both learned from The People's Salon that music is a shared responsibility of musicians and audience.

## Conclusion: Working Together

The four vignettes from our fieldwork exemplify the many conversations that shaped The People's Salon. In the process of making a concert experiment happen, we had to learn and develop the skills that are needed to co-create a musical event and, also, to develop mutual understanding. We had to "learn how to rehearse cooperation, exploring its different forms" (Sennett, 2012, p. 24). This was all but easy. As any large organization, a symphony orchestra has to follow certain logics—of planning, scheduling, and realizing artistic quality—that limited the time and space for creative and open conversations needed to come to unexpected results and insights. And, as academic researchers, our styles of reasoning often failed to resonate with how the orchestral practitioners framed their work experiences and goals. Having to realize concrete products within a certain time frame frequently led to dialectic conversations where common ground took the form of pragmatic solutions. These differences also explain that the overall project had various specific outcomes. For the orchestra, The People's Salon gave them a new concert format that can be repeated, as currently happens in fact under the Covid-19 related restrictions. Stories about classical music told by Friends were recorded on video, and some of these served as an introduction to a streamed concert.

For the researchers, doing the collaborative experiments resulted in practical and theoretical insights that are shared with relevant scientific communities through publications and presentations, as in this chapter.

Coming back to the promise of interdisciplinary collaboration outlined at the start of this chapter, namely that it will bring innovation, we feel that our project should have been characterized by more sustained dialogical conversations. That we were able to produce results together indicates that our dialectical conversations were successful. We did share a commitment to the outcome, and sympathy allowed us, at least imaginatively, to identify with the other actors we worked with. In the everyday practice of doing research together, however, it was difficult to find or organize moments to empathize, to leave the safety of our routines and self-definitions, and to really wonder how and why others work the way they do. This is where dialogic conversations have an open-ended character: their goal is not consensus, but learning through being curious about the other. We realize that the conversational ideal types we borrow from Sennett cannot do justice to the complexity of all the things that happened, but they do help to draw lessons from our project that may be helpful to others who collaborate to fulfil the promise of innovation and change. Collaboration is a skill that does not only take time, but that also needs care, imagination, and the willingness to experience a sense of surprise. Instead of working toward the closure of collective results, it aims at the open-endedness of continued learning from each other. Organizing this learning is a matter not only of scheduling meetings but also of truly having an interest in what working together may bring, and in the skills needed in Sennett's rehearsal: musicking and communicating dialogically.

# References

Artful Participation. (2021, December 1). www.artfulparticipation.nl

Bonds, M. E. (2014). *Absolute music: The history of an idea.* Oxford University Press.

Born, G. (2010). For a relational musicology: Music and interdisciplinarity, beyond the practice turn. The 2007 Dent Medal Address. *Journal of the Royal Musical Association, 135*(2), 205–243. https://doi.org/10.1080/02690403. 2010.506265

Born, G., & Barry, A. (2014). Interdisciplinarity: Reconfigurations of the social and natural sciences. In A. Barry & G. Born (Eds.), *Interdisciplinarity. Reconfigurations of the social and natural sciences* (pp. 1–56). Routledge.

Burland, K., & Pitts, S. (2016). *Coughing and clapping: Investigating audience experience*. Routledge.

"Call for Proposals Creative Industries: Smart Culture—Arts and Culture Call" (2016). NWO.

Cook, N. (2014). *Beyond the score: Music as performance*. Oxford University Press.

Dewey, J. (2005). *Art as experience*. Penguin. (Original work published 1934).

Eve, I. (2019, June). Project design The People's Salon. [Unpublished project document].

Eve, I. (2020). *The same but differently*. Zuyd University/Lectoraat AOK.

Gadamer, H.-G. (1989). *Truth and method*. Continuum. (Original work published 1960).

Goehr, L. (1992). *The imaginary museum of musical works: An essay in the philosophy of music*. Oxford University Press.

Hennion, A. (2015). *The passion for music: A sociology of mediation* (M. Rigaud & P. Collier, Trans.). Ashgate.

Idema, J. (2012). *Present! Rethinking live classical music*. Muziek Centrum Nederland.

Marres, N. (2012). Experiment: The experiment in living. In C. Lury & N. Wakeford (Eds.), *Inventive methods: The happening of the social* (pp. 76–95). Routledge.

Ministerie van Onderwijs, Cultuur en Wetenschap. (2011). *Meer dan kwaliteit: Een nieuwe visie op cultuurbeleid*. Ministerie van OCW.

Ministerie van Onderwijs, Cultuur en Wetenschap. (2013). *Cultuur beweeg: De betekenis van cultuur in een veranderende samenleving*. Ministerie van OCW.

Peters, P. F. (2019). *Unfinished symphonies* [Inaugural lecture]. Datawyse/Universitaire Pers Maastricht.

Pinch, T., & Bijsterveld, K. (2003). "Should one applaud?" Breaches and boundaries in the reception of new technology in music. *Technology and culture, 44*(3), 536–559.

Raad voor Cultuur. (2014). *De cultuurverkenning: Ontwikkelingen en trends in het culturele leven in Nederland*. Raad voor Cultuur.

Sennett, R. (2012). *Together: The rituals, pleasures and politics of cooperation*. Allen Lane.

Small, C. (1998). *Musicking: The meanings of performing and listening*. Wesleyan University Press.

Smithuijsen, C. (2001). *Een verbazende stilte: Klassieke muziek, gedragsregels en sociale controle in de concertzaal*. Boekmanstudies.

Toelle, J., & Sloboda, J. (2019). *The audience as artist?* The audience's experience of participatory music. *Musicae Scientiae, 25*(1). https://doi.org/10.1177/1029864919844804

# Alignment and Alienation: Emergency Staff and Midwifery Scholars as Co-researchers

Jessica Mesman

## Introduction

Complex practices, such as the emergency department or intensive care unit in a hospital, the control-room for train traffic or air transportation, or the kitchen of a busy restaurant, are highly dynamic and usually clear manifestations of ingenuity and responsiveness. Scholars in the interdisciplinary field of Science and Technology Studies (STS) try to unravel such practices by acknowledging the relevance of what is traditionally ignored or considered unimportant. As an STS scholar, I thus pay close attention to the role of the 'mundane' in stabilizing practices and I explore whether its unpacking may help us to learn from it. More often than not, ordinary daily work proves to involve plenty of rich and resourceful actions and activities, rather

I would like to thank Katherine Carroll and Irene Korstjens for our valuable partnership.

J. Mesman (✉)
Department of Society Studies, Faculty of Arts & Social Sciences,
Maastricht University, Maastricht, The Netherlands
e-mail: j.mesman@maastrichtuniversity.nl

**273**

K. Bijsterveld, A. Swinnen (eds.), *Interdisciplinarity in the Scholarly Life Cycle*,
https://doi.org/10.1007/978-3-031-11108-2_15

than merely routine ones. I identify informal processes as the glue that holds practices together. This counter-intuitive way of reasoning also makes me stress the potential of 'ambivalence' and 'inconsistency' in preventing practices from stalling. In sum, by considering the 'ordinary' as an extraordinary accomplishment, it becomes possible to question why things go well and what is the role of the attributes that are usually neglected when trying to raise our understanding of complex practices.

Unpacking common, day-to-day matters is important not only from an academic point of view but also for practice optimalization. Although innovation is the 'traditional' way to improve practices, in line with Rein De Wilde (2000), I take an exnovative approach. According to De Wilde, innovation makes us blind to the importance of what is already in place. Practice improvement, he argues, also requires 'exnovation': the explication of the hidden strength of practices and to learn from this. After all, the achievement of quality may in part be the product of an unplanned yet effective set of initiatives. By exposing what is already there, exnovation acknowledges that unarticulated actions serve as a vital resource for the accomplishment of work and opportunities for improvement. Scrutinizing practices from such an exnovative angle may generate input for both scholarly work and practice optimization.

Everyday application or recurrent use of solutions within practices, however, is likely to turn effective routines into habitual forms of conduct that are hardly noticed anymore. Although present, practitioners no longer pay attention to them. How, then, are we as researchers to identify these potential resources for improvement when practitioners themselves may no longer be aware of them? In order to gain access to what is taken-for-granted, we potentially benefit from finding ways to combine an outsider's awareness with an insider's understanding. It is here that video-reflexive ethnography (VRE) comes in, as this approach allows the familiar and unfamiliar to coincide, as it were.

Video-reflexive ethnography is a visual, collaborative, and interventionist method for studying practices by video-recording day-to-day work and analyzing this footage together with the practitioners in reflexive sessions (Iedema et al., 2019). In these sessions, practitioners have an opportunity to look at their own workplace from a different angle. This new way of perceiving their daily routines, in combination with the 'outsider-questions' posed by the researchers, has the potential to exnovate the 'hidden' strengths of practices. The collaborative analysis of the footage opens room for identification and clarification, awareness, and appreciation, sharing experiences, questioning assumptions, and discussing suggestions for improvement. These analytic reflections are recorded and subsequently used for further analysis by the researchers. This twofold analysis provides the input for both scholarly output and practice optimalization. Importantly, such collaboration with participants is not limited to the analysis of footage in the reflexive meetings but runs all the way from agenda-setting, what to film (as well as when and where), the selection of clips for the reflexive discussions, and their analysis in the reflexivity sessions to publishing the ultimate results (e.g., Carroll et al., 2021). In VRE, research subjects truly act as co-researchers. As researcher, my focus on the taken-for-granted within practices necessitates interdisciplinary collaboration because it is impossible for me to unpack 'the everyday' without those who inhabit it. The VRE method provides the formula to do so.

This perfect fit between the VRE method and my ambitions to unravel complex practices and making a difference in these practices hardly guarantees a smooth, easy ride. As I study predominantly medical practices, my co-researchers do not only have a different disciplinary background, but they also rely on an epistemic culture (a realist paradigm) that differs from mine (constructivist paradigm). The epistemic culture in medicine is dominated by the idea that the biomedical realities involved are objectively observable and exist independently of the human knower. With adequate methods, science can provide an objective description of those realities. This epistemic orientation contrasts with the constructivist paradigm serving as my epistemic base, which defines realities as being socially and experientially based and, as such, allows for multiple realities. In other words, VRE projects in healthcare are not only interdisciplinary but also multi-paradigmatic.

Although the paradigmatic underpinning of research projects tends to reside in the background, occasionally, it will take up a foreground position. In this chapter, I will use two case studies to discuss one issue in particular: how paradigmatic differences impact trust in professional credibility in the context of interdisciplinary collaboration—trust in epistemic qualities is key in interdisciplinary research. By means of the case studies, I will discuss how interdisciplinary collaboration has the potential to undermine epistemic credibility when acting as a relativist in a realist environment. The first case study is based on a project conducted together with Katherine Carroll, a VRE colleague, in an Emergency Department in an Australian hospital. The argument of this case study draws on one of our publications (Mesman & Carroll, 2021) in which we demonstrate how a solid preparation of research includes potential risks for the professional reputation of the involved researchers and as such for building interdisciplinary alliances. While the focus of the first case study is on establishing collaboration with research subjects, the second case study shifts the attention to collaborative dynamics within the research team itself. Based on a project in Midwifery Science in the Netherlands, this case study allows me to explain how real or imagined threats to professional credibility can affect the experience of interdisciplinary collaboration, as well as the relationship with one's own academic community. By discussing and comparing both case studies, I aim to convey situations that generate potential risks to researchers for loss of their epistemic credibility, as well as opportunities for on-the-ground problem-solving aimed at preserving or reclaiming their professional reputation.

## Methodological Rigor Comes with Mess

### Vignette: First Impressions

> Emergency Department (ED) ward: Entering the ED, we are welcomed by the chief medical specialist. He shows us around on the ED and introduces us to several ED staff members. We see nurses attending to documentation and talking to patients. Curtains around patients are being closed and opened again, and while we hear phones ring, medical staff is constantly walking in and out, attending to their tasks. This orientation on the ward gives us a first impression of what this particular ED is like. (Adapted from Carroll & Mesman, 2011, p. 160)

The focus of our project in the Australian ED was on the ways in which clinicians—including ambulance personnel, triage nurses, and emergency doctors—hand over patient information effectively within such a complex work environment. By identifying their collaborative strengths, we aimed to contribute to improving clinical handovers in the ED. To unravel their styles and strategies, we used the method of video-reflexive ethnography. However, for VRE to work and result in reflexive learning and intervening, ED staff had to become co-researchers. Therefore, we had to get them motivated to join us in the first place. Clearly, we had to do some groundwork first.

Like any other ethnographic research, building a methodological foundation for applying VRE involves more than getting approvals and explaining your project in boardrooms and staff meetings. First and foremost, it requires trust and learning the cues of the practice (Carroll & Mesman, 2011; Iedema et al., 2019). Positioning themselves in the local ecology of the work site enables researchers to build rapport and to get acquainted with the different roles, responsibilities, and organizational rhythms and structures. Such a contextual exploration also helps to identify key-informants, as well as the best moments to film and the right spots to be in. In addition, being among staff all day allows researchers to further explain the project's focus and set-up on a more individual basis, and, most importantly, to build trust to recruit clinicians to become engaged as co-researcher.

Fully aware of the importance of a solid preparation, Katherine and I went into the Emergency ward once we felt ready to do so. Quite soon, however, things became rather challenging, and, unexpectedly, at one point, even our professional reputation was at risk.

## Vignette: Finding the Right Spot

ED ward: Being donned in scrubs, we feel more than ready to start filming clinical handovers. We are informed by the clinical staff what time and location their handovers are usually done in the ED. To be able to capture them on film, we had to find a place that will allow us to do so without disturbing their handover meetings. In consultation with the ED staff, we position ourselves next to the pole in the middle of the ward's workstation with our camera ready. (Adapted from Carroll & Mesman, 2011, pp. 161–162)

As detailed in this vignette, we decided on the best location for filming the handover practices based on information of the ED staff. It turned out, however, that clinicians engage in many informal handover activities as well. These ad hoc exchanges were also vital input for our project because they constituted a substantial segment of the ED handover practice. Yet, these informal handovers, as we found, were not always neatly planned or organized. In fact, many of them were quite discreet, involving multiple informal handovers concurrently or taking place at various locations at unpredictable moments. To capture the informal handovers on film, though, we had to able to recognize them as such. This turned out to be rather difficult.

## Vignette: Finding the Right Conversation

> ED ward: 'Come on, let's go over there. They are doing one, I'm sure'! Afraid to miss specific handover moments, we leave our position next to the pole and move around to be able to film as many as possible encounters between staff members. This results in us franticly filming all kinds of information exchanges. After all, it might involve a handover activity. As we try to get a sense of the various handover practices, the clinicians can see us literally running around with a camera, filming whoever is talking, and apologizing for recording the wrong conversation or for being too late for the right ones. As a result, they look at us clearly wondering what on earth we are doing. (Adapted from Carroll & Mesman, 2011, p. 162)

Unlike in other research traditions, such as when doing laboratory experiments, in the case of video-reflexive ethnographic studies, methodological thoroughness is not defined and applied *before* you enter the field site. Instead, it is generated on location as the activities to be studied unfold (Iedema et al., 2019). To build a solid research infrastructure in such an evolving context implies a dramatic change for the position of researchers, because all efforts to develop a sound methodological foundation occur in real time, in front of the eyes of everyone present. If all goes well, this is not a problem. Frequently, however, this stage will also be marred by uncertainties, if not outright mistakes. This is part and parcel not only of the VRE methodology but also of research in general.

Despite descriptions of orderly research processes in the methodology sections of much academic literature, all research involves a rather messy process to some extent, as argued by John Law (2004). From this perspective, us 'running around' with a camera was just a dynamic form of getting to know the place. Our asking 'stupid' questions or filming the wrong conversations should be considered to some extent as an inevitable, 'messy' dimension of the preparatory stage of the research involved.

The usual dynamic for a consultative and collaborative research method, in other words, will involve a disorderly process. It is in the interrelated turbulence that researchers will learn about and connect with others. The fact that we, as ethnographers, are at ease with our messy preparation does not guarantee that the same applies to others as well. In an environment like healthcare, where common sense realism is the dominant framework, it will be harder, we argued, for (prospective) research participants to accept disorderly actions as part of serious and solid research (Mesman & Carroll, 2021). Healthcare staff is familiar with a pre-fixed, coherent research protocol with a well-formulated hypothesis to be tested on the basis of pre-defined conditions and steps to be taken. Instead, they saw two women running around with a camera, filming every conversation, including those about Saturday-evening plans. The contrast between the messiness of our preparatory work and their experiences with the clean set-up of medicine's random controlled trials harbors the risk of us being regarded as probably lost, presumably methodologically inadequate. While VRE groundwork is aimed at relationship building and starting up collaboration, its local and immediate character is potentially harmful to the participants' responsiveness and the researchers' trustworthiness. In other words, the actions required to build a solid foundation for collaboration can simultaneously jeopardize its realization.

How, then, can researchers ensure being taken seriously and maintain participants' interest in collaborating actively in their project? Our VRE project on the ED survived the perils of exposure and we succeeded to enlist staff as our co-researchers. For one thing, it was the hierarchical culture of medical practice that opened up possibilities for safeguarding our professional reputation (Carroll & Mesman, 2011). In the hierarchical ecology of the ED, our academic seniority and institutional affiliations buttressed our credibility as researchers. It positioned us on a level that gained us the attention of the chief clinicians involved. Their support was key to be accepted by their colleagues as well. The fact that we were seen with the chief clinician on the first day secured our position. Despite we could read amazement on the faces of some of the ED staff about our way of doing research, being seen in the presence of their clinical management caused them to assume everything was okay. Furthermore, their being re-assured made them more approachable for us, allowing us to interact with them and to build respectful relationships. These interactions also provided possibilities for further explanation of our ethnographic project and its methods, which contributed to the staff's understanding of our ways of doing and reasoning. Over time, they too considered our 'messiness' as 'constructive adjustment' and 'required flexibility' (Mesman & Carroll 2021, p. 172). In retrospect, doing our preparatory work in the 'here and now' had the potential of creating confusion or even suspicion about the quality of our project or the credentials of us as researchers. But this full exposure also provided us the possibility to establish recognition because our 'messy' approach was accompanied by a 'relationally-driven, collaborative, transparent, humble and therefore trustworthy' attitude (Mesman & Carroll, 2021, p. 172).

This case study demonstrates how paradigmatic differences can impact interdisciplinary alliances in unforeseen ways. Evidently, learning the cues and recruiting staff simultaneously in the same space creates potential risks for the professional reputation of researchers. At the same time, being 'on location' can also provide the means to maintain or re-establish professional reliability before damage is done beyond repair. Yet, the potential for undermining a researcher's position is not limited to the interaction with research subjects. Within a research team, paradigmatic diversity can also cause concern about the professional credibility. My next case study will display how collaboration across disciplines comes with expertise and ignorance, and how these can impact a researcher's professional reputation.

# Professional Credibility Meets Professional Identity

According to Kim Fortun and Todd Cherkasky, collaboration is all about 'diversity' in which we understand 'diversity as a resource' (1998, p. 146). People with different expertise align into a synchronized effort to accomplish something that could not be done otherwise. In this way 'collaboration marks the difference between those who work together rather than their sameness' (Fortun & Cherkasky, 1998, p. 146). Interdisciplinary collaboration takes diversity to another level. The set-up of interdisciplinary collaboration is affected by methodologies and theoretical choices, as linked to specific tasks, discourses, modes of practice, roles, and responsibilities. Methodological and theoretical decisions are framed by, for example, the distribution of power, paradigmatic (in)compatibilities, and requirements of the field of application. In this way, interdisciplinary collaboration also implies feelings of being more or being less 'at home.' The potentially uncertain role of 'visiting researchers' who venture beyond their disciplinary comfort zone comes with insecurity that resembles fieldwork-related anxieties as described by Jörg Niewöhner (2016): 'will they like me?' (relationship building), 'will they tolerate me?' (epistemic authority), and 'will I have something new to add?' (research output). My second case study, from research into Maternity Care, serves as basis for discussing such insecurities in relation to interdisciplinary collaboration. This discussion has its focus on the concern of losing one's epistemic authority and shows how solving this problem comes with a risk of drifting away too far from one's own disciplinary turf.

Maternity Care aims for a safe birth trajectory for mother and child. Considering the complexities involved in everyday Maternity Care practices, a group of scholars in Midwifery Science initiated a study on the ways midwives, obstetricians, and parents-to-be accomplish effective collaboration (Korstjens et al., 2021). Aiming at in-depth insights into the implicit ways professionals establish constructive interactions among themselves and with parents, video-reflexive ethnography was applied as methodological approach. Because of my VRE expertise, I was invited to join the research team, which I consider a privilege. Because of the team's intellectual drive and know-how of the maternity field, the whole trajectory was a highly valuable experience for me. But occasionally, I felt also confused and ambivalent about our project (Smolka & Mesman, forthcoming). In some cases, I even wondered about the status of my epistemic credibility and felt alienated amidst my midwifery colleagues. Such feelings of uncertainty also deserve attention when reflecting on interdisciplinary collaboration. Based on my experiences in the midwifery team, I will unpack situations that generate doubt about epistemic authority, specifically in relation to my own insecurity.

## Management of Ignorance

In many disciplines, including Midwifery Science, it is common practice to perform a systematic review of existing theories on the topic of investigation before doing the actual empirical research. The aim of such a systematic search is to identify the knowledge already available in the literature and to develop sensitizing concepts that directs the empirical data collection and analysis. The midwifery team aimed to get the results of this review published in a Maternity Care journal that had a strong focus on quantitative research. Hence, for strategic reasons, our literature search was partly based on a quantitative approach, which took me to unknown places. In VRE, the principle of ethnographic openness and the input of the co-researchers rule out the pre-fixed focus of a literature review. Having been a VRE researcher for over a decade, a preliminary literature search was no longer part of my way of working when I started my collaboration with the midwifery scholars.

## Vignette: Getting Lost and Losing Face

> Midwifery Department (MD): the principal investigator (PI) of the project, a PhD student, and myself are sitting in an office of the MD. We discuss our work on the literature review. The PI explains what actions to take to ensure 'content validation,' 'trustworthiness,' and 'credibility.' She is clearly on familiar grounds. I, on the other hand, am struggling and, too many times, I wonder what she is talking about. Some of the words she

uses and the actions she proposes are anything but familiar to me. But I don't dare to ask at this moment. Not again. I am sure she must be rather fed up with my questions. Sometimes, I sense frustration on her part, given that all these questions delay the research process. It is evident that my questions surprise her. Not because they are so brilliant but because they are so basic. Sometimes, I see disbelief in her eyes: 'What do you mean, you don't know'? More than ever, I am aware of my ignorance in this area. I realize they invited me as an expert, but in too many moments I act like a novice. I am out of my depth and feel stupid and embarrassed. I clearly sense a dissonance between the expectations and my performance. I fear this impression will damage my scholarly reputation and my position in the team. After all, how many questions are you allowed to ask before losing your credibility as an expert?

Not being familiar with the selected methodologies or theoretical armature is a common situation in interdisciplinary research (e.g., Fitzgerald & Callard, 2015). After all, intellectual pluralism is a leading motive for interdisciplinary collaboration. Yet, capitalizing on complementary knowledge implies at least some level of ignorance on the part of those involved. Moreover, according to Maria Jönsson and Anna Rådström (2013), what we don't know says something important about who we are. Because ignorance is the automatic result of being disciplined, it is not something to be embarrassed about. Likewise, Zachary Piso et al. (2016) have argued that while you acquire the knowledge characteristic of your field, you are also being disciplined in which aspects to ignore because they are deemed irrelevant. This distribution of attention implies that ignorance is actively produced while gaining expertise in your field. In this way, interdisciplinary research reinforces ignorance on an individual level, making us aware of being a 'situated knower' (Piso et al., 2016, p. 648).

At the same time, being formally excused for lack of knowledge by being disciplined does not take away the potential impact of (false) expectations on your professional reputation. In the midwifery project, I was expected to have expertise not only in VRE research but also—if implicitly—in performing such a systematic literature search. Considering my background in qualitative research, it was assumed by my midwifery colleagues that a systematic review was a standard technique in my research practice as well. Certainly, I know my way around in scholarly literature, but I lack the expertise to perform a review based on their standards and (partly) quantitative approach. Moreover, as VRE scholar, I was 'disciplined' to ignore any pre-defining activities, including systematic reviews aiming to define sensitizing concepts that would guide our data collection and analysis. In this project, however, Midwifery Science featured as the dominant disciplinary orientation. Accordingly, the entire set-up was based on its paradigmatic understanding of doing qualitative research. Furthermore, the communicative context of the project, that is, Maternity Care journals, needed additional methodological steps to be taken to meet the field's research quality criteria. But this required expertise was a lacuna on my part. Obviously, a mismatch between expectations and performance can potentially weaken one's epistemic authority. Even though the interactions of the team did not give any indication of such a problem, I grew concerned that my academic reputation would be undermined before taking up my responsibility as VRE expert in the second stage of the project in which the VRE method was center stage.

## Paradigmatic Convictions Under Pressure

Being afraid that my lack of expertise in the literature review stage of the project would negatively affect my expert position in the VRE stage, I had to re-secure my professional credibility. To regain the team's trust (if I had lost it at all), I decided to ignore my uncertainties and instead open up, learning new skills, and go for it by joining them wholeheartedly. Although everything worked out well, my enthusiasm also came with costs.

## Vignette: Full Immersion and Lost Again

MD: To my own surprise, I hear myself discussing issues of bias with my midwifery colleagues and join them in actions aimed at content validation. We work together on the basis of strategies to accomplish these aims by analyzing data together, alternated with independent assessments based on strictly worded selection criteria. Checks and balances include performing duplicate extractions from systematic collected samples, and discussions until consensus is reached. We read and re-read all verbatim transcripts, developed a coding scheme, and defined categories. We chart huge tables, make flowcharts, and make sure readers can evaluate whether our findings are transferable to other contexts. Clearly, methodological data and investigator triangulation are high on our agenda. (Adapted from Helmond van et al., 2015, pp. 211–212)

For many researchers, most of these actions will look familiar. After all, this is how research is done. Although being impressed by our work, I also felt a mixture of alienation and awkwardness. The terms mentioned were not quite my language. Nor was the cross-coding my way of reasoning and ordering data. In other words, I did not recognize my usual 'ethnographic me' in this project. Leaving one's academic turf raises the question of how far one can travel and spend time somewhere else before getting lost. To answer this question requires more insight into the underlying causes of my somewhat disturbed academic self. While being fully immersed in the project, I became concerned about the fact that I did not only join a team with a different disciplinary profile but that I also had to work within a different paradigmatic tradition. Paradigmatic differences imply that researchers will navigate other standards, tasks, or requirements for doing research. VRE is anchored in a post-qualitative tradition (Iedema et al., 2019). This implies that VRE privileges being over knowing, style over method, entanglements and multiplicities over binaries and categories, emergence and fluidity over stability and fixedness, engagement over professional distance, and creating to enable over thinking to know, just to name a few characteristics. Indeed, my colleagues from Midwifery Science looked at some things from the other extreme of these various binary opposites. By joining them I was crossing a line.

Making use of VRE for over a decade caused me to internalize its research style. Therefore, it should come as no surprise that while being engrossed in a highly systematic exploration of the literature to predetermine the focus of our ethnographic part of the project, the VRE scholar in me resisted. Such ways of doing and thinking are, moreover, at odds with my identity as an STS scholar. However, when I joined the team, I did so without much thought about these concerns. As all team members were qualitative researchers who agreed on using VRE as a main research instrument, I didn't bother about spelling out our paradigmatic positions. After all, all of us were delighted with each other's expertise. Looking back, it is easy to be surprised about this naïveté, but one should also realize that, in many research projects, an eagerness to get going may push such more fundamental considerations toward the background (e.g., Fitzgerald et al., 2014).

To secure my position in the midwifery team, I had committed myself to their ways of doing research. However, joining them with enthusiasm in activities that were aimed at meeting the quality standards of their field merely relocated—rather than solved—my concern of losing my academic credibility. And, this time, it was related to my own discipline. Many of my activities were based on assumptions rejected by me and most of my colleagues. Triangulation, for example, assumes that there is a fixed point of reality at which all perspectives meet. Avoiding bias, to name another example, aims for neutrality. Both these considerations do not align with the overall epistemic framework on the basis of which I conduct research. I felt that I had drifted too far away from my scholarly moorings, and I began to fear that my STS/VRE colleagues would come across a publication with my name on it that presented a methodological argument from which we as STS scholars, as a matter of principle, aim to move away. I felt like a vegetarian about to get caught in a steakhouse. All kinds of thoughts crossed my mind: would such a publication give the impression that my epistemic convictions can change overnight? Should my name be on the final version? Was I disloyal to my own community? Was it self-betrayal to my own epistemic convictions?

How to stay true to one's own paradigmatic convictions while being fully submerged in interdisciplinary teamwork based on other research principles? One strategy is to perceive the activities of the interdisciplinary team as a topic of investigation as well (e.g., Fitzgerald et al., 2014; Haapasaari et al., 2012). Because STS scholars are interested in knowledge production, membership of a multi-paradigmatic team provides a splendid opportunity for doing an epistemic ethnography. When involved in activities that are geared toward credibility, trustworthiness, and triangulation, researchers find themselves in an excellent position to study knowledge production 'from within.' In such a situation, gaining new competences will not only serve the main research project but also re-align researchers with their own academic community: in my case, the STS community for whom analyzing knowledge production is an important building block of the field. Furthermore, studying the activities of the team also creates a platform for deliberation, as other team members have their own moments of confusion and surprise while having a stranger in their midst.

Looking back makes me realize that it was 'difference' that created discomfort. It was difference that made me feel insecure as well as forced me to choose sides. It was also difference that made me feel alienated and that gave rise to my study of the epistemic culture of my collaborators. In other words, interdisciplinary collaboration kept me busy trying to resolve differences. Evidently, there is a lot to learn from studies on inter-disciplinarity in practice, where a focus on collaborative unity is replaced by 'careful equivocation' (Yates-Doerr, 2019). Instead of trying to find the same scale of valuation, a common referent, or reconcile tensions, it is argued that researchers should honor differences and learn from it (e.g., Fitzgerald et al., 2014; Yates-Doerr, 2019). In such a frame, feelings of disloyalty are absent, for no one has to choose sides, as difference is not a binary logic but a mode of productive relating (Yates-Doerr, 2019). In sum, interdisciplinary collaboration implies not only specific expertise but also specific ignorance. As portrayed in the case study, ignorance can generate insecurity about the status of one's own epistemic authority. One way to deal with this is to close the knowledge gap and resolve the difference. Joining the midwifery team in their analysis and learning the ropes in combination with my own epistemic ethnography was my strat-egy to build bridges to each side. However, other studies on interdisci-plinary collaboration offer a different view. In this perspective there is no gap as ignorance is just the other side of the coin and, as such, uncer-tainty over epistemic reputation resolves. Next time I 'go native,' I travel the trails of ignorance with my head held high.

## Concluding Remarks

When in the early 2000s, I embraced exnovation as a productive concept that motivated me to develop video-reflexive ethnography as a fertile analytical approach, this provided a strong boost to my interdisciplinary profile and focus. Over time, I experienced how differences between disciplinary conventions can produce tensions, feelings of estrangement, and uncertainty. These effects point out how repertoires of reasoning, normative orders of justification, vocabularies of communication, and scripts of doing are at the heart of our professional identity. The two case studies presented here demonstrate the rich potential of interdisciplinary research. For one, such research will open doors to new field sites and journals that used to be out of reach. Moreover, joining an interdisciplinary team may result in a profound learning experience, leading to a more detailed awareness and higher appreciation of other ways of collecting and analyzing data. It is here that we touch upon one of the major advantages of doing interdisciplinary research: it allows researchers to move from 'being acquainted with' to 'having know-how' regarding other research traditions.

To be sure, both case studies also underscore how moving around in other paradigmatic landscapes can create discomfort and real or imagined threats for professional credibility. Paradigmatic discrepancies will challenge our positions and arguments, and this exposes our research principles and processes not only to others but also to ourselves. Being experienced researchers, we may think to know them all; over time, we may have become unaware of many of them. In this way, interdisciplinary teamwork is exnovative in itself: instead of a camera perspective, it is the paradigmatic contrast that provides a window for looking at our own research practice from another angle and learn from it.

While doing STS requires a relativist perspective as to the theories and methods the knowledge practices under study convey, studying exnovation with VRE both challenges and articulates the abilities we need for acting as a relativist in a realist environment. To effectively adopt pluralism as a strategic choice and to feel comfortable in doing so requires a specific set of attitudes and actions (Fortun & Cherkasky, 1998). First, it calls for intellectual outward looking and internal permissiveness that encourage an openness to each other's way of reasoning, including its norms and values (Castree et al., 2009). Also important is the art of attentiveness that prompts one to listen carefully and pay respectful attention to get to know each other and learn how to craft meaningful responses (van Dooren et al., 2016). Such attention is vital for responding appropriately in a research process based on diversity. These terms of engagement will benefit from a reflexive monitoring of the research dynamics. Consequently, working together on how to work together is the indispensable dialog that underpins interdisciplinary research. Building bridges in this way allows for a mutual exploration. This learning process involves a journey through another methodological and conceptual landscape. These new experiences in doing solid research in other traditions will impress, make one humble, and—most significantly—prevent dogmatism.

# References

Carroll, K., Mesman, J., McLeod, H., Boughey, J., Keeney, G., & Habermann, E. (2021). Seeing what works: Identifying and enhancing successful interprofessional collaboration between pathology and surgery. *Journal of Interprofessional Care, 35*(4), 490–502. https://doi.org/10.1080/13561820.2018.1536041

Carroll, K., & Mesman, J. (2011). Ethnographic context meets ethnographic biography: A challenge for the mores of doing fieldwork. *International Journal of Multiple Research Approaches, 5*(2), 155–168.

Castree, K., Demeritt, D., & Liverman, D. (2009). Introduction: Making sense of environmental geography. In K. Castree, D. Demeritt, D. Liverman, & B. Rhods (Eds.), *A Companion of Environmental Geography* (pp. 1–15). Wiley-Blackwell.

Dooren van, T., Kirksey, E., & Münster, U. (2016). Multispecies studies: Cultivating arts of attentiveness. *Environmental Humanities, 8*(1), 1–23.

Fitzgerald, D., & Callard, F. (2015). Social science and neuroscience beyond interdisciplinarity: Experimental entanglements. *Theory, Culture & Society, 32*(1), 3–32. https://doi.org/10.1177/0263276414537319

Fitzgerald, D., Littlefield, M. M., Knudsen, K. J., Tonks, J., & Dietz, M. J. (2014). Ambivalence, equivocation and the politics of experimental knowledge: A transdisciplinary neuroscience encounter. *Social Studies of Science, 44*(5), 701–721. https://doi.org/10.1177/0306312714531473

Fortun, K., & Cherkasky, T. (1998). Counter-expertise and the politics of collaboration. *Science as Culture, 7*(2), 145–172. https://doi.org/10.1080/09505439809526499

Haapasaari, P., Kulmala, S., & Kuikka, S. (2012). Growing into interdisciplinarity: How to converge biology, economics, and social science in fishery research? *Ecology and Society, 17*(1), 1–12. https://doi.org/10.5751/ES-04503-170106

Helmond van, I., Korstjens, I., Mesman, J., Nieuwenhuijze, M., Horstman, K., Scheepers, H., Spaanderman, M., Keulen, J., & De Vries, R. (2015). What makes for good collaboration and communication in maternity care? A scoping study. *International Journal of Childbirth, 5*(4), 210–225. https://doi.org/10.1891/2156-5287.5.4.210

Iedema, R., Carroll, K., Collier, A., Hor, S., Mesman, J., & Wyer, M. (2019). *Video-reflexive ethnography in health research and healthcare improvement: Theory and application.* CRC Press.

Jönsson, M., & Rådström, A. (2013). Experiences of research collaboration in 'soloist' disciplines: On the importance of not knowing and learning from affects of shame, ambivalence, and insecurity. In G. Griffin, A. Bränström-Ohman, & H. Kulman (Eds.), *The emotional politics of research collaboration* (pp. 130–143). Routledge.

Korstjens, I., Mesman, J., DeVries, R., & Nieuwenhuijze, M. (2021). The paradoxes of communication and collaboration in maternity care: A video-reflexivity study with professionals and parents. *Women and Birth, 34*(2), 145–153. https://doi.org/10.1016/j.wombi.2020.01.014

Law, J. (2004). *After method: Mess in social science research.* Routledge.

Mesman, J., & Carroll, K. (2021). The art of staying with making & doing: Exnovating video-reflexive ethnography. In G. Downey & T. Zuiderent-Jerak (Eds.), *Making and doing: Activating STS through knowledge expression and travel* (pp. 155–177). The MIT Press. https://doi.org/10.7551/mitpress/11310.001.0001

Niewöhner, J. (2016). Co-laborative anthropology: Crafting reflexivities experimentally. In J. Jouhki, & T. Steel (Eds.), *Etnologinen tulkinta ja analyysi. Kohti avoimempaa tutkimusprosessia* [Ethnological interpretation and analysis: Towards a transparent research process]. (pp. 81–125). Ethnos.

Piso, Z., Sertler, E., Malavisi, A., Marable, K., Jensen, E., Gonnerman, C., & O'Rourke, M. (2016). The production and reinforcement of ignorance in collaborative interdisciplinary research. *Social Epistemology, 30*(5–6), 643–664. https://doi.org/10.1080/02691728.2016.1213328

Smolka, M., & Mesman, J. (forthcoming). Practicing care-as-affect and engagement-as-critique: Careful engagement with video-reflexive ethnography and socio-technical integration research. [Manuscript in preparation]. In D. Lydahl & N. C. Nickelsen (Eds.), *Ethical and methodological dilemmas in social science interventions - careful engagements in healthcare, museums, design and beyond.* Springer.

Wilde, R. (2000, December 2). Innovating innovation: A contribution to the philosophy of the future. International Conference *Policy agendas for Sustainable Technological Innovation* (POSTI). London.

Yates-Doerr, E. (2019). Whose global, which health? Unsettling collaboration with careful equivocation. *American Anthropologist, 121*(2), 297–310. https://doi.org/10.1111/aman.13259

# 'Doing' Teamwork as 'Doing' Family: Researching Transnational Migrant Families Through Interdisciplinary Collaboration

Valentina Mazzucato, Bilisuma Dito, and Karlijn Haagsman

## Introduction

Teamwork is frequently discussed as being integral to interdisciplinary research, but it does not just happen. It takes work. This chapter focuses on the work that is behind teamwork and especially the less-discussed elements relating to the personal and emotional work that it entails. The authors are three researchers who have worked together since 2010 on two large interdisciplinary and international research projects on transnational families. These projects study how people 'do' family across international borders. Families are not just a connection of people related through blood or marriage ties, but they are defined and held together

V. Mazzucato (✉) • B. Dito • K. Haagsman
Department of Society Studies, Faculty of Arts & Social Sciences,
Maastricht University, Maastricht, The Netherlands
e-mail: v.mazzucato@maastrichtuniversity.nl;
bilisuma.dito@maastrichtuniversity.nl; r.haagsman@maastrichtuniversity.nl

© The Author(s) 2023
K. Bijsterveld, A. Swinnen (eds.), *Interdisciplinarity in the Scholarly Life Cycle*,
https://doi.org/10.1007/978-3-031-11108-2_16

through the 'work' that goes into giving and receiving care to its members. This work entails commitments and obligations, supported by communication, trust, and constructive frictions that keep families acting as such. In this chapter, we draw parallels between 'doing' family and 'doing' teamwork by focusing on these elements—communication, trust, and working with frictions—to unpack the work that is entailed in 'doing' teamwork. This chapter is structured as an open conversation between the three researchers. We chose this format as an illustration of the way we work together. Conversation is an integral part of the praxis of working together that we developed over the years. Inclusive conversation allows for exploration and unexpected insights to come to the fore, and it contributes to the creation of a team in which members feel heard and respected.

A note on the interdisciplinary projects: The Transnational Child Raising Arrangements projects (TCRA and TCRAf-Eu), funded by the Dutch Research Council (NWO) and the New Opportunities for Research Funding Agency Co-operation in Europe (NORFACE), respectively, studied transnational families in which parents migrate from sub-Saharan Africa to Europe and some or all their children remain in the country of origin in the care of someone else. The main aim was to investigate the effects of geographical separation on child raising practices across borders. We followed migrant parents in Europe and their children and caregivers in Africa. These projects included researchers from anthropology, migration studies, child psychology, development economics, demography, geography, political science, and family sociology. We employed both qualitative ethnographic and quantitative survey methods. Mazzucato was the Principal Investigator of the projects, and Dito and Haagsman were post-doc and PhD, respectively.

# Open Communication

Key for transnational child caring arrangements to work is communication. There is a lot of 'work' in communicating: for migrant parents to negotiate their needs in a new country with those of their child and the caregiver; for the caregiver in the country of origin to manage the information flow to the parent in order not to worry them; and for the child to vent their frustrations to the parent while catering to the caregiver's demands. All of these situations entail daily communication and negotiation between actors, in the form of small, everyday decisions, such as when to make a phone call, what to say on the phone, what to leave out, how to communicate a matter, whose needs take priority, and how to come to a common solution so that the transnational family continues to function as such (Poeze et al., 2017). The same is true for the importance of communication in teamwork, involving everyday communications, decisions of when and how to communicate, all of which foster a team feeling. As with transnational families, this section will show that interdisciplinary collaboration requires constant communication and dialogue, regular meetings, curiosity, and an open mind.

VALENTINA: In heading interdisciplinary research projects, I pay a lot of attention to how to establish good communication. This has to do with meeting frequently at the beginning of a project, also just to get to know each other. It is often said about interdisciplinary collaboration that you need to work on establishing some ground rules and developing a common language as disciplines come with their own theories and concepts. But to me, what is even more important is the atmosphere that one establishes: it needs to be open, people need to feel that they can contribute, no matter what their perspective is. There needs to be respect for each other's methods and, most of all, people need to feel safe being vulnerable. That is, that they can say when they are unsure about something, or admit that they may have made a mistake, or are free to explore something when they don't yet know where it will lead them. To create such an atmosphere, you need to meet frequently, for work, but also include some other kinds of meetings. For example, I instituted a yearly three-day research retreat with only one rule: if it is not fun, we don't do it. During these retreats we talk about our research but in very different ways than we are used to, such as through making a theater play about it or having workshops on making people feel heard, seen, and respected (Liberating Structures, n.d.). I think these moments are fundamental in creating trust so that, when we have work meetings, people are respectful of and trust each other.

KARLIJN: I agree, and a lot happens before you actually get started with the data collection. It's helpful to be clear about things from the start. For example, one needs to talk about the praxis of collecting and sharing data before the data collection starts as there are many ways to do this. We also met on a weekly basis to develop a methodology and read literature to create a common conceptual framework. I remember that at first, as a PhD on the project, I was quite shy. But the weekly meetings and the team activities really helped me to feel comfortable to speak up, even if it was to say that I didn't understand something. Teamwork is not something you just jump into; there's really a lot of preparatory work, and it takes time, and a lot centers around communication.

BILISUMA: For me, one of the fundamental aspects of interdisciplinary teamwork and having good communication is the attitude of the team members: they need to be open-minded. Understanding the limitations of one's own training is a good way to start, at least that was true for me. My interdisciplinary PhD environment really helped me to integrate within the team of researchers in the TCRA project who came from different disciplinary backgrounds. Even though I was one of the few quantitative scholars on the team, I was met with an open-mindedness by team members that paved the way for developing collaborative working relationships. For me, this was fundamental for the articles that I co-authored: the theoretical framework for the quantitative analyses that I did, incorporated the insights from the qualitative researchers to identify the variables that we included in our models. The reason why I could shift my path of inquiry and include new kinds of variables, was that I had these variables at my disposal because we discussed, as a team, the different elements that we felt might underlie parental choices to migrate without their children (Dito et al., 2017), which we drew from our different disciplinary knowledge. So, I ended up including variables that I would not have thought of on my own, drawing only from my discipline of economics. This was possible because of the interdisciplinary questionnaire designed by other team members before I joined the project.

KARLIJN: Also during the interpretation of results, it was really important to communicate in the team and be open to other interpretations. I remember when we found that the sex of migrant parents did not make a difference in terms of their well-being outcomes. We could have stopped there. Yet, we asked the ethnographer on the team (Miranda Poeze) to look into this. She found that women and men do experience different hurdles, so her work allowed us to understand our own results better. Combining our two results, we came to our interpretation that men experience and talk about separation in a different way, yet, in terms of well-being measures, they are impacted as much as women (Haagsman et al., 2015). This was a new finding in the literature.

BILISUMA: Yes, the whole is more than the sum of its parts. To achieve this, especially in integrated projects such as ours, there's also the need to share data. But we also found that data sharing was not always easy. When team members are less enthusiastic to share their data, it can limit team collaborations. Sometimes this is inevitable, especially when people come from disciplines where they are not trained to work in teams. But you can go a long way to minimizing this if teams engage openly and continuously about the possibilities and the limits of data sharing. I also think co-authorship among team members with different disciplinary and methodological backgrounds can facilitate data sharing since it creates a sense of joint ownership of the end product. But I guess we'll return to this point later in the conversation.

## Trust

One of the things that made transnational families function well, where members, although separated by thousands of kilometers, were still able to maintain a sense of family and well-being, was trust (Haagsman et al., 2015). Parents had to trust that the caregivers in the origin country were taking good care of their children (Poeze et al., 2017), caregivers, in turn, had to trust that parents were doing their best, even when remittances were not forthcoming, and children had to trust that their parents were interested in their well-being even when parents were not able to send money home (Dankyi et al., 2016). What created this trust? Being vulnerable, giving each other the benefit of the doubt and taking time. These are elements we also find in teamwork. Yet, contrary to families, teams do not start with a shared history. They therefore first have to work on building a community of practice within which trust can be built.

KARLIJN: What I said earlier about daring to speak up as a PhD student in a team of people from different disciplines but also in different academic positions has to do with trust—trust that I won't be judged if I dare to be vulnerable. For example, at the beginning, when I had to present my work in our team meetings, I was quite nervous, because I was not yet sure of what direction I wanted to go in. But because we met frequently, and we all shared our unfinished work, we also shared our insecurities. Valentina set the tone by showing her own doubts and creating a sense that we were all in the same boat. We were all quite supportive of each other. This didn't mean we avoided difficult discussions. In fact, we had to be willing to sometimes go back to the drawing board, such as when Miranda (the ethnographer) pointed out to me that I should nuance the interpretation of my results, or when Bilisuma (the economist) pointed out that I had missed some important variables in my statistical models.

BILISUMA: Yes, it also feels uncomfortable sometimes. I remember feeling a bit hesitant about some of our quantitative models because we were not preoccupied with methodological issues that economists obsess about such as endogeneity. But as I worked more in the project, I realized that, at the end of the day, we were not out to prove to the world that our models were perfect, but rather we were filling a gap in the literature with unique data and interdisciplinary insights that would help to further the field of transnational family research.

Sometimes, with my quantitative methods, I inevitably need to simplify reality. Then, I get so many detailed questions from the ethnographers, asking me to delve into nuances. But this is quantitative work, how am I supposed to incorporate all these ideas? So, my first reaction was to become defensive. But, then, because of the constructive way the feedback was given, I realized I needed to be careful with my generalizations, to qualify things. You know, I come from a tradition of a very structured way of writing research papers. It is a different way of presenting data than the more storytelling way of qualitative researchers. At first, I felt very uncomfortable with this. I would think, okay, how am I supposed to make sense of all this fuzziness. Remember Karlijn, we'd have many conversations about this. I sometimes just didn't understand what the ethnographers were trying to bring to the table.

KARLIJN: Yeah, it's funny, I had the opposite experience because I originally studied anthropology and in my first quantitative papers, my supervisors told me to be less descriptive and more to the point. And remember Bilisuma when we first co-authored together? You had a certain way of writing up the analysis that was so different from mine, which we struggled with in the beginning, but through practice and being open about this difficulty we merged our two ways and found a common way.

VALENTINA: I think a key to teamwork is to be okay with feeling uncomfortable, knowing that your teammates won't judge you for mistakes or not knowing. I draw inspiration from David Bohm, a physicist who has thought long and hard about creative dialogue. He states, 'communication can lead to the creation of something new only if people are able freely to listen to each other, without prejudice, and without trying to influence each other' (1996, p. 3) (Fig. 1). And this stance of 'we're all in this together to come to a greater understanding' really requires time. It takes time to build trust, through many conversations, to build a common vision. Because working together is also about daring to change your own mindset and that takes time.

**Fig. 1** Visualization of 'creating something new together' © Mazzucato et al.

In the TCRA projects we questioned the notion of family that is prevalent in family sociology and child development psychology. We developed more dynamic categories of families that included important members at a distance, who these members are, and how these compositions change over time (Mazzucato & Dito, 2018). But developing these categories was a whole process of thinking that evolved over time and in conversation with other researchers on the team who kept questioning us. Having this questioning attitude helps you become reflexive about the way you are used to seeing things. It's so easy to fall into tunnel vision just because we are used to thinking of a particular societal issue, or a group of people, in a certain way. We inherit certain categories, as it were, from our disciplines. We all use categories—even people who are critical of categories, use categories. We can't get away from using them, but we can question the categories we use, reflect on the kind of knowledge they generate, and be critical of them by proposing alternative or additional categories. This is where a real added value of teamwork comes in.

BILISUMA: This makes me think of the inter-cultural dimension of teams. It is not per se that you have to have people from different cultures to do interdisciplinary teamwork, but, just like disciplines can create tunnel vision, so can having people all from the same cultural background. I remember that in the TCRA project, we had discussions about child raising norms in the context of Ghana and comparing them with Asian contexts, where most other studies were done. I remember feeling slightly uncomfortable with saying 'culture in Ghana is like this.' I think we managed to a certain extent to not make such big generalizations about cultural norms, but it was important to have this continuous reflection about what culture means. We had some heated discussions, I remember. New generations have different norms than older generations, and urban inhabitants different from rural ones. So there are differences even within a culture.

KARLIJN: Yes, I remember a discussion I had with a Dutch woman at a conference. Even though we talk about Dutch families as nuclear, the extended family is important for child raising. Just look at the role of grandparents in taking care of their grandchildren in The Netherlands.

# Frictions

Family relationships are about commitments and obligations. Sometimes such commitments are not easy to fulfill and can lead to frictions. Migrant parents might find it hard to send money home because of their precarious positions in the country where they migrated to. Caregivers who care for migrants' children in the origin country may see their economic situation waning, making it more difficult to care for the children. Children may also find it difficult to focus on their schoolwork without the emotional support of their parents. Those transnational families that we found functioned best, in terms of fulfilling the emotional and care needs of family members, were those that, despite the frictions, were able to oblige the commitments to each other (Poeze et al., 2017). Also in interdisciplinary teamwork, frictions can be experienced either within the team or with the broader academic world that is still dominantly organized along disciplines. The point is not to avoid frictions but to learn how best to deal with them in a team.

VALENTINA: One thing that is often said about interdisciplinary research is that it is hard to get published because most journals are still disciplinary. How do you feel about this?

KARLIJN: Well, I think it depends on which journal, I guess. Luckily, in the field of migration, there are more interdisciplinary journals. I think we always chose the journals that were open to this kind of interdisciplinary research, also as we wanted to reach academics working in multiple fields and hence wanted to make sure we did not only reach one discipline. We put our topic and findings central rather than the discipline.

VALENTINA: That's true, but don't forget that we also published in psychology journals showing that if you have something innovative to say to disciplinary theories, sometimes, it is even possible to publish in disciplinary journals.

KARLIJN: True, but we also experienced some frictions with disciplinary scholars and this was quite challenging to deal with, emotionally. I remember when I presented our findings at a conference, I was attacked by an anthropologist. The simple fact that I presented numbers made her discredit the results. At another conference, I was in a panel with mainly economists, and they criticized my analysis for not following a certain procedure. I was a PhD student and these attacks felt very personal. It's important to develop thick skin. It really helped to have the project team with whom to discuss the criticisms and realize that part of the criticism was because we were innovating—doing things out of what is considered 'normal.' As the team members were from various disciplines, they could help me understand the feedback and how to deal with it. They helped me put it in perspective. This gave me confidence.

BILISUMA: I have a question for you Valentina. I think the issues we raised before about openness or vulnerability and building trust in teams, also have to do with personality. Maybe it is important not to have a big ego? In these projects, you selected the team members. How do you go about selecting a team? What do you look for when you are hiring people?

VALENTINA: Well, aside looking for competent people, I also look for collaborators. So, during job interviews, I ask about people's prior experience with teamwork, what they liked but also what they found challenging, I ask about concrete experiences with teamwork that did not function, because this happens a lot. What's important for me is not that they did not have negative experiences but what explanations and lessons they draw from such experiences. Also, if someone has no or very few co-authored articles, then this is something I ask about. Yes, and indeed, it is best to avoid big egos.

KARLIJN: Oh, I remember this one candidate we were interviewing for a position on the team. They had a fantastic CV and grades. They also gave great answers to our questions. But then I made a mistake because I had overseen something on their CV and the way they reacted, offended, and really blaming me for the mistake, do you remember Valentina?

VALENTINA: Yes, it was the best thing you could have done, making that mistake because their true character came out. I thought to myself, oooh, if someone reacts like this for a simple mistake, this does not promise well for collaboration.

BILISUMA: I think what's also important is the complementary nature of our knowledge. I mean, I was the only economist on the team, but instead of feeling isolated I felt I could actually contribute to this team because the team members appreciated my insights. There was this feeling of 'my contributions matter.' Valentina, you've always been the PI on these projects, but I never felt the hierarchy in conceptual or analytical discussions. That is, you have your strong opinions on many issues, but personally, I never felt that you didn't take my inputs on board. I actually felt respected as a researcher. And I think this is key to interdisciplinary collaboration: we need to learn from each other. If it is only the PI's or the professor's ideas that count, then you don't learn from each other.

KARLIJN: In a way, this points to the importance of listening, truly listening, and showing empathy to the people on the team.

VALENTINA: It makes me feel happy to hear you say this. Because, of course, this is part of the challenge as well in teamwork: there needs to be someone who makes sure the project stays on course and delivers what it had promised to the funder. On the other hand, you want team members to feel like they have the freedom to make the project theirs. This can lead to frictions, although I have to say that, when this happened, it was really important to be able to explain to the person involved what our obligations are, as a team, toward the funder and what we promise. It helps if you have the time in projects to really build up a common vision of what the aims of the project are and how each of us has a role in achieving these aims.

One thing that contributes to the coherence and cohesiveness of projects is co-authorship. That is, the idea that we will all show up in the outputs of the project also creates more of a sense that we are in this together and it eases the hurdles that team members may feel about sharing 'their' data. Data are not owned by the person who collects them. Data are the product of a whole team effort that starts from the preliminary team meetings aimed at creating a common vision, to establishing the methodology, to developing the tools, developing an analytical frame, analyzing the data, and writing up the findings. Too often in the social sciences and humanities, research is attributed to the writing phase. But there are so many phases that lead to the actual final written product and so many people involved in the different phases. The hard sciences such as Medicine or Biology are much better at acknowledging research as a collective effort. I think true interdisciplinary collaborations should always strive for co-authored outputs—it's a way to acknowledge everyone along the process.

BILISUMA: On co-authorship, I think that the length of our projects helped us. We always worked on longer projects of about four years which gave us the time and opportunity for more cross-pollinated research papers. But what is key is that these expectations and different ambitions of the team members and the PI are discussed at the start of the project and get revisited during the project.

KARLIJN: Yes, and co-authorship is still not always appreciated. For example, in selection committees of funding agencies or academic positions, sometimes applicants are judged by how many single-authored papers they have produced. It seems like they think that co-authored papers require less effort. I find this problematic, as we have been saying, to work together actually costs a lot of work and leads to better and more nuanced insights. I know for sure that the papers I've co-authored are better than if I had done everything by myself. They are more nuanced, innovative, and integrate more literature. So, to then say you are not a good researcher because you mainly co-produced, is narrow-minded and simply wrong. But this is mainly academics from particular disciplines which are not used to co-authorship.

BILISUMA: That's an important point that needs to be raised because co-authorships are for sure results of interdisciplinary collaborations. Your point hits the mark on what kinds of attitudes and cultures need to change in academia. We cannot encourage interdisciplinary research without changing the reward and recognition systems in place.

VALENTINA: Indeed, but the system is also us, and it is by actually engaging in interdisciplinary projects, appreciating what we gain from them, while also tackling the difficulties, that we can change a research system to one that acknowledges that much research is, at its essence, a collaborative endeavor.

Well, as always, it's been fun and insightful to engage in these reflections with you both!

# Conclusion

This dialogue was the result of a series of recorded unstructured conversations about our experiences working in interdisciplinary projects together. We relistened to the conversations and clustered the points around three themes that resulted as most prominent: open communication, trust among team members, and the ability to constructively deal with frictions. We realized that these themes paralleled those of our research on transnational families, which, with hindsight, is not surprising. In both our research topic and in our teamwork we pay attention to the work that is entailed in making collaboration function. Although one can argue that these themes are important in all teamwork, not just interdisciplinary teamwork, they are even more important in interdisciplinary teams because the potential for miscommunication, lack of trust, and discord is greater when one comes from different academic disciplines, each with its own vocabulary and praxis. Just like in transnational families, the distance makes family 'work' even more essential when one cannot count on everyday physical exchanges.

Our work has benefitted from our interdisciplinary collaboration. However, interdisciplinary teamwork is not easy and doesn't come automatically. Interdisciplinary teamwork takes time and is also an emotional investment that not everyone is willing to make. It necessitates empathy and reflexivity. The atmosphere needs to be one of open communication and trust so that members can dare to be vulnerable and share unfinished work. Open-mindedness is necessary to be willing to call into question one's own ways. While everyone needs to feel free to share ideas, we benefitted from clear hierarchy where one PI steered the team in one direction but also ensured that conversations continued and that data are shared and published in a fair and inclusive way. In sum, interdisciplinary teamwork is made up of small, everyday actions that are often invisible or go unnoticed but which, together, build up to a praxis and an atmosphere conducive to collaboration and interconnection.

# References

Bohm, D. (1996). *On dialogue*. Routledge.

Dankyi, E., Mazzucato, V., & Manuh, T. (2016). Global social protection as a reciprocal process: The work of caregivers in providing care for migrants' children. *Oxford Development Studies, 45*(1), 80–95. https://doi.org/10.1080/13600818.2015.1124078

Dito, B., Mazzucato, V., & Schans, D. (2017). The effects of transnational parenting on the subjective health and well-being of Ghanaian migrants in The Netherlands. *Population, Space and Place, 23*(3), e2006. https://doi.org/10.1002/psp.2006

Haagsman, K., Mazzucato, V., & Dito, B. B. (2015). Transnational families and the subjective well-being of migrant parents: Angolan and Nigerian parents in the Netherlands. *Ethnic and Racial Studies, 38*(15), 2652–2671. https://doi.org/10.1080/01419870.2015.1037783

Liberating Structures. (n.d.). *Liberating structures. Including and unleashing everyone*. Retrieved November 29, 2021, from www.liberatingstructures.org

Mazzucato, V., & Dito, B. (2018). Transnational families: Cross-country comparative perspectives. *Population Space and Place, 24*(7), 2165. https://doi.org/10.1002/psp.2165

Poeze, M., Dankyi, E. K., & Mazzucato, V. (2017). Navigating transnational childcare relationships: Migrant parents and their children's caregivers in the origin country. *Global Networks, 17*(1), 111–129. https://doi.org/10.1111/glob.12135

# Author Index[1]

---

[1] Note: Page numbers followed by 'n' refer to notes.

© The Author(s) 2023
K. Bijsterveld, A. Swinnen (eds.), *Interdisciplinarity in the Scholarly Life Cycle*,
https://doi.org/10.1007/978-3-031-11108-2

# Subject Index[1]

---

[1] Note: Page numbers followed by 'n' refer to notes.

© The Author(s) 2023
K. Bijsterveld, A. Swinnen (eds.), *Interdisciplinarity in the Scholarly Life Cycle*,
https://doi.org/10.1007/978-3-031-11108-2